TECHNOLOGY, INNOVATION and POLICY 4

Series of the Fraunhofer Institute
for Systems and Innovation Research (ISI)

A study on behalf of

VDI/VDE-Technologiezentrum
Informationstechnik GmbH
Rheinstr. 10 B

D-14513 Teltow, Germany

and

The German Marshall Fund of the United States
Friedrichstr. 113a

D-10117 Berlin, Germany

Head Office:
11 Dupont Circle, N.W.
Washington D.C. 20036
USA

Oliver Pfirrmann · Udo Wupperfeld
Joshua Lerner

Venture Capital and New Technology Based Firms

An US-German Comparison

With 1 Figure
and 18 Tables

Physica-Verlag
A Springer-Verlag Company

Dr. Oliver Pfirrmann
Freie Universität Berlin, Arbeitsstelle Politik und Technik
Ihnestr. 53, D-14195 Berlin, Germany

Professor Dr. Udo Wupperfeld
Hofecker 7, D-68782 Brühl, Germany

Professor Dr. Joshua Lerner
Harvard University, Graduate School of Business Administration
Morgan Hall, Soldiers Field
Boston, Massachusetts 02163, USA

ISBN-13: 978-3-7908-0968-8   e-ISBN-13: 978-3-642-48683-8
DOI: 10.1007/978-3-642-48683-8

Die Deutsche Bibliothek – CIP-Einheitsaufnahme
Pfirrmann, Oliver:
Venture capital and new technology based firms : an US-
German comparison; with 18 tables; [a study] / Oliver
Pfirrmann; Udo Wupperfeld; Joshua Lerner. [On behalf of
VDI/VDE-Technologiezentrum Informationstechnik GmbH
and The German Marshall Fund of the United States]. –
Heidelberg : Physica-Verl., 1997
  (Technology, innovation, and policy; 4)
  ISBN-13: 978-3-7908-0968-8
NE: Wupperfeld; Udo:; Lerner, Josgua:; GT

Cover design: Erich Kirchner, Heidelberg
SPIN 10549373   88/2202-5 4 3 2 1 0-Printed on acid-free paper

Preface

The following publication has been taken from the internal final report that was prepared for the research project "The Role of Venture Capital in the Formation and Growth of New Technology Based Firms: An US-German Comparison". The research project was supported by the German Marshall Fund within the framework of the program "Grants for Research on the US or Transatlantic Relations"(Grant No. A-2043-02). The project started on April 1, 1995 and finished on March 31, 1996. The research team consisted of three researchers from Germany and the US. They are presented in alphabetical order:

-       Professor Joshua Lerner, Harvard University, Graduate School of Business Administration, Boston,

-       Dr. Oliver Pfirrmann, Freie Universität Berlin, Arbeitsstelle Politik und Technik,

-       Prof. Dr. Udo Wupperfeld, Fraunhofer-Institut für Systemtechnik und Innovationsforschung, (FhG-ISI), Karlsruhe, Fachhochschule Pforzheim.

The research team also gratefully acknowledges the support of Dr. Gordon Murray from Warwick Business School, the University of Warwick, England. Dr. Murray provided informational support and helpful comments on the outcome of the research team's work.

During the project work, discussions were held with German experts from the banking and VC-sector. All of the meetings took place in Berlin. Several meetings were held with Mr. Roger Bendisch of the Seed Capital Fund (SCF) and with Dr.

Alexander Schrader from Deutsche Bank Research, Bayerische Vereinsbank. The research team gratefully acknowledges the support given by these experts in providing information.

At the end of the research project, on March 7, 1996, a one-day workshop was held in Berlin with twenty-six experts representing venture capital, industry, politics and academic institutions. The research team presented selected results under the heading "Perspectives for Venture Capital in Germany - Results from an US-German Study". The lectures held at the workshop also included statements given by venture capitalists, managers of NTBFs and policy-makers. The lectures were discussed intensively, and the some of the major findings are included in the overall conclusions.

The research team would like to thank all other people and institutions, especially VDI/VDE-Technologiezentrum Informationstechnik GmbH that also supported the project. However the research team takes full responsibility for any remaining errors.

Oliver Pfirrmann   Udo Wupperfeld   Joshua Lerner

Table of Contents   Page

List of Tables and Figures

Glossary

ACMC	Aberlyn Capital Management Company
ADW	Akademie der Wissenschaften der DDR
AMEX	American Stock Exchange
AKU	Allgemeine Kapitalunion GmbH & Co. KG
ARD	American Research and Development
BAYTEP	Bayerisches Technologieentwicklungsprogramm
BJTU	Modellversuch „Beteiligungskapital für junge Technologie-unternehmen"
BMW	Bayerische Motorenwerke
BTU	Modellversuch „Beteiligungskapital für kleine Technologieunternehmen"
BVK	Bundesverband Deutscher Kapitalbeteiligungsgesellschaften e.V.
c.	circa
CEO	Chief Economic Officer
CFO	Chief Finance Officer
CRADA	Cooperative Research and Development Agreement
DBG	Deutsche Beteiligungsgesellschaft m.b.H.
DEC	Digital Equipment Corporation
DM	Deutsche Mark
DtA	Deutsche Ausgleichsbank
EKH	Eigenkapitalhilfe - Programm
ERP	European Recovery Programme
EstG	Einkommenssteuergesetz
ETH	European Technologieholding N. V.
EU	Europäische Union
EVCA	European Venture Capital Association
FhG-ISI	Fraunhofer-Institut für Systemtechnik und Innovationsforschung
FLIP	Finance Lease on Intellectual Property
FU Berlin	Freie Universität Berlin

GeBeKa	Gesellschaft für Beteiligungen und Kapitalverwaltung m.b.H. & Co.
GmbH	Gesellschaft mit beschränkter Haftung
HLT	Hessische Landesentwicklungs- und Treuhandgesellschaft m.b.H
IPO	Initial Public Offering
IRR	Internal Rate of Return
ISC	Illinois Superconductor Corporation
IVCP	International Venture Capital Partners S.A. Holding
JTU	Junges Technologieunternehmen
KBG	Kapitalbeteiligungsgesellschaft m.b.H.
KDV	Kapitalbeteiligungsgesellschaft der Deutschen Versicherungswirtschaft
KfW	Kreditanstalt für Wiederaufbau
KG	Kommanditgesellschaft
LBB	Landesbank Berlin
LBO	Leveraged buy-out
MBG	Mittelständische Beteiligungsgesellschaft
MBI	Management-buy-in
MBO	Management-buy-out
MIT	Massachusetts Institute of Technology
NASDAQ	National Association of Securities Dealers Automated Quotation System
NBL	Neue Bundesländer
NMB	Niederländische Mittelstandsbank
NMS	National Market System
NTBF	New Technology Based Firm
NYSE	New York Stock Exchange
OEM	Original Equipment Manufacturer
OHG	Offene Handelsgesellschaft
OTC	Over The Counter
R&D	Research and Development
SBA	Small Business Administration
SBIC	Small Business Investment Company

SBIR	Small Business Innovation Research Program
SCC	Seed Capital Companies
SCF	Seed Capital Fund
SEC	Securities and Exchange Commission
Tbg	Technologiebeteiligungsgesellschaft
TIG	Technologie-Investitionsgesellschaft
TOU	Modellversuch „Technologieorientierte Unternehmensgründungen"
TVM	Techno Venture Management GmbH & Co. KG
US	United States of America
u.a.	und andere
VAG	Versicherungsaufsichtsgesetz
VC	Venture Capital
VCC	Venture Capital Companies
WFG	Deutsche Gesellschaft für Wagniskapital m.b.H
XTV	Xerox Technology Ventures

1 Introduction

1.1 Objectives and the Scope of the Study

The objectives of this comparative study are to generate analytical insights into the role of venture capital as a means of financing new technology based firms. Venture capital does not solely focus on new technology based firms. For example, financing the expansion of established companies as well as the financing of management buyouts and buyins constitutes a major part of this investment business. However, with regard to the historical role VC has played in the development of today's large businesses such as Apple, Advanced Micro Devices, Digital Equipment, or Intel, new technology based firms can be seen as important vehicle in the commercialization of technological inventions. It is well-known that the process of developing new technology involves uncertainty and risk. Hence, venture capital - in its traditional role - is a suitable means of fostering technological development.[1]

However, the nature of investment in technological inventions is complex. All the necessary expenditure involved in an innovation project can be regarded as a single investment with a long-term return, depending on the commercial success of the resulting product, service or process. There is a basic conflict in that such expenditure is very diverse, some being devoted to the traditional aspects of investment (machinery, etc.), whereas other expenditures are regarded by

---

[1] For a comprehensive overview about NTBF see for example: Kulicke u.a. 1993, Chancen und Risiken junger Technologieunternehmen - Ergebnisse des Modellversuchs "Förderung technologieorientierter Unternehmensgründungen (TOU), Heidelberg; for an excellent state-of-the-art discussion about venture capital: Bygrave and Timmons, 1992, Venture Capital at the Crossroads, Boston

conventional accounting as current expenditures (R&D, training, marketing, etc.). The attempt to categorize these different types of investment as well as the investment of tangible assets in intangible assets deserves mention in this context.

Traditional financiers are obviously going to focus on material investments and operations that produce tradable assets. Only specialist investors will venture into operations with a high intangible component and/or where the specific assets of the enterprise play a key role.

Small companies are inherently at a disadvantage in capital markets, since they are individually subject to greater risks and because investigation costs for external investors, as a proportion of the amount invested, are inevitably greater the smaller the firm. These problems are compounded by the natural reluctance of entrepreneurs to allow the outside interference that external equity involves. Against this background, venture capital assumes a very important role both for the financing of product development as well as marketing and for promoting the expansion of an enterprise. Venture capitalists take an equity stake in the firms they finance, sharing both the upside and downside risks. Most of the firms that receive venture capital financing are unlikely candidates for alternative sources of funding. They have few tangible assets to pledge as collateral, and produce operating losses for many years.

Meanwhile external investors are cautious about locking up funds in companies from which there is no easy way out. The use of retained earnings, asset-based finance including bank loans and direct investments, as in the case of business angels for example, are all means of helping companies expand to the point at which it becomes lucrative for venture capitalists to participate in the financing of a small firm. Venture capitalists will be forced to rely on trade sales to large companies if there is too great a discrepancy between the bottom end on public securities markets and the level at which realizations, or even partial realizations, are necessary for them to turnover their capital quickly enough to satisfy their own investors. However, these difficulties do not always apply to the financing of management buyouts and buyins of larger groups, where the risk is easier to assess and divested companies tend to be relatively large.

It is well known that small companies have few opportunities to gain access to stock markets. These imperfections in the capital market with respect to supplying finance to smaller firms are primarily seen in the inability of external investors to evaluate the quality of investment opportunities.[2] This even applies to smaller companies quoted on main and secondary markets. Recent research has shown that investments made by smaller quoted companies tend to vary with cash flow, whereas large companies, owing to their better access to financing, are able to invest independently of their own cash flows.[3]

The financing of new technology based firms (NTBFs) is easier in one respect, but in others it is more difficult than is generally the case with smaller firms. It is easier because both the large amounts of capital frequently required and the shorter time window available for exploiting new technological developments mean that entrepreneurs are more willing to seek external equity finance. It is more difficult to finance NTBFs because risks are not easy to assess, given both the nature of innovative technologies and the typical, pressing need of NTBFs to compete on a global scale. A long period of time may also pass between the research and development phase and the prospect of a positive cash flow. This is particularly true of certain fields of specialized technology, such as biotechnology. It is, however, also particularly true of the specific framework conditions for NTBFs, for example, the markets and the supportive infrastructure.

---

[2]   Cf. Pratten 1993

[3]   Cf. Fazzari et al. 1988

1.2   Issues, Research Questions and the Study Concept

Issues and Research Questions

Past examples of today's large businesses from the US demonstrate that, in some cases, venture capital can have a significant impact on NTBFs. In general, venture capitalists are active investors. They monitor the progress of firms, sit on boards of directors, and mete out financing based on the attainment of milestones. Venture capitalists retain the right to appoint key managers and remove members from the entrepreneurial team. In addition, venture capitalists provide entrepreneurs with management support, e.g. access to consultants, investment bankers and lawyers.

However, while the differences between the US and German VC markets seem to have diminished over time, there nevertheless still remain several marked differences between the US and German venture capital industries.

Firstly, the size and rate of growth of these funds is very different. The US is by far the largest venture pool, with over 34 billion dollars in invested capital. The German volume is considerably smaller with roughly DM 6.2 billion in invested capital. Secondly, the composition of investments differs considerably, too. Investments in buyouts were far more common in the US, while seed and early-stage investments were noted more frequently in Germany. A third area in which a difference is apparent is that of the funding sources. Banks play a more significant role in Germany, whereas the US pension funds were more substantial investors in the US than they were in Germany. Finally, public promotion in Germany has generated a larger supply of venture capital there than in the US.

Another significant set of differences is to be seen in the legal structure of venture capital funds. Most recent venture funds formed in the US were structured as limited partnerships, i.e. composed of juristic persons, one for management and one for investment. In Germany, very different structures predominant. For instance, the fund structure most frequently encountered in Germany is the "general partnership", i.e. one juristic person, frequently a bank, is responsible for both management and investment. In fact, private banks are generally only involved with financial and not with management support. The structures available for venture investments are very unattractive for foreign investors.

Thirdly, the market structure of the VC industries is different in both countries. In the US, venture capital firms were responsible for virtually all the professional private equity investment that took place during the 1960s. Over the subsequent decades, the sector split into distinct industries involved in buy-out, "special situation," and start-up investments. Despite these divisions, many venture funds did make at least a few buy-out investments during the 1980s, and some funds continued to invest in a variety of different asset classes. Whereas the US industry arose in the 1940s, a venture capital industry cannot be said to have emerged in Germany until the early 1980s. Despite the many governmental and private activities pursued since the mid-eighties, especially with respect to NTBFs, the German VC-market is now lagging behind in the mid-nineties.

Thus the analysis will provide answers to the following questions:

- What are the framework conditions for VC industries in the US and Germany?
- How does the entire private equity market in each country influence the availability of venture capital?
- Are there different performance strategies between the US and German venture capitalists vis-à-vis their portfolios, and what kinds of strategies can be identified?
- Are there quantitative and qualitative differences with respect to the involvement of venture capitalists in new technology based firms in Germany and the US?

The Concept of the Study

The concept of investigation was based on a three-stage approach. For the general tasks of the study the research team used existing literature, statistical sources and data bases as well as official publications, for example, those of the national venture capital organizations. Based on an evaluation of the literature available, hypotheses had to be elaborated for ten case studies on NTBFs with venture capital in the US and Germany. In both countries, especially in the US, there is an extensive body of literature, both theoretical and empirical in nature, including some case studies.[4] However, as far as the issues of the analysis are concerned, an evaluation of the literature did not seem to be sufficient owing to the lack of (current) comparative US-German analyses.[5] We therefore decided not to undertake a broad empirical investigation, but to conduct case studies in order to qualify the results of the literature review and to provide empirical insights into the central questions of the study. Thus the investigation approach is more a qualitative/exploratory in nature.[6]

The case studies included personal interviews with the entrepreneur(s) of NTBFs and the relevant venture capital firm(s). The interviews were based on a guideline providing open questions in order to achieve a broad description of the development and situation of an NTBF and the strategy of a VC company. The results of the literature review and the empirical work were used to prepare a workshop with experts from VC companies as well as academics and entrepreneurs from the US and Germany. The aim of this was to reflect upon the results and to draw a conclusive picture of the relationship between NTBFs and VC in both countries.

---

[4] Cf. Albach et al. 1986; Dean/Giglierano 1989; Frear/Wetzel 1990; Gillner 1984; Gompers 1994; Gompers/Lerner 1994; Lerner 1994 a/b, 1995; Klemm 1988; Lampe 1992; Picot et al. 1989; Sapienza et al. 1992; Sapienza/Timmons 1989 a/b

[5] Cf. Workshop 1983, Riesenhuber 1984; an interesting comparative analysis for UK and Germany is provided by Gerybadze 1988.

[6] See for discussion of methodological aspects of explorative research design chapter 4

Following the introduction (chapter 1 of the report), chapter 2 gives a description and definition of the area of investigation. First of all, the different segments of the US and German capital market are discussed in light of their relevance to NTBFs. Subsequently a detailed outline of our understanding of NTBFs is presented alongside a description of possible financial links to VC.

In chapter 3 the results of the literature screening are presented. The research team compiled a large stock of literature consisting of scientific journals and books as well as internal reports that were published officially and non-officially (for example by the European Commission, national venture capital associations, governmental agencies and other universities). The comparison of the framework conditions for venture capital in the US and Germany is presented as a summary of our findings from the literature for each country in the following chapters. The first part will consider the development and structure of the venture capital market. It will include certain topics such as its historical development, a description of the market today and a characterization of the types of venture capital companies active on the VC market in each country. The following part will consider the framework conditions and the development determinants specific to each country. This analysis includes such issues as taxation and legal aspects, the possibilities of performing disinvestments, the role of other capital suppliers and financing sources as well as the legal investment requirements for institutional investors. A specific chapter will be devoted to policies to stimulate VC markets.

The results of our empirical investigation are presented in chapter 4. The first part deals with the methodological aspects of the investigation. Different approaches to case study investigation were discussed within the research team. After considering the Harvard Business School approach, which requires several months of detailed investigation in each case, and the German approach, which envisages shorter investigation periods, the research team decided to examine the opportunities and limits of a one year research project. Instead of elaborating a few case studies intensively, the empirical section presents ten case studies based on the German approach. At the same time, the study was able to deal more adequately with the heterogeneous German venture capital scene, which displays at least three different forms of venture capital investment. The two sections that follow present five case studies for both countries. Each case study begins with a profile of the NTBF and describes the VC investment from the standpoint of the portfolio company. The specific characteristics of the investment company are described here, e.g. fund structures, partners, areas of activity and management strategies. The case study finishes with a detailed description of the relationship between the VC company and the NTBF.

Certain kinds of information could not have been obtained from the empirical investigations. With regard to the efficiency of the VC companies, the IRR[7] provides a common standard for evaluating the results of investments. However, in the majority of the case studies, even in Germany, IRRs were not made public to third parties or proved to be inapplicable because they only showed a small number of exits. Thus the IRR was not taken into account in the empirical section. Nevertheless, chapter 3 does contain statements about the IRR on US funds. In the case of NTBFs, however, it was not possible - in either country - to match groups of firms that correspond to one another in almost all relevant company characteristics, e.g. age, size, sector/technology, relationship to investment company. For an analysis of the impact of investments on the development of enterprises, in terms of employment growth for example, too many variables had to be controlled for side effects. Furthermore, the illustrative character of theses studies only permits tentative conclusions about the impact of venture capital on the development of NTBFs.

Chapter 5 contains a summary of the results and the conclusions of the research.

---

7   The IRR - the internal rate of return - is quantified in a spreadsheet in which future cash streams are discounted back to a net zero figure after investment and other costs.

2 Venture Capital Markets, New Technology Based Firms and Financial Systems: Definitions and an Overview of the US and German Experiences

2.1 Venture Capital: the Process and Investors

There are many different definitions of what is meant by venture capital. As late as the mid- seventies, Liles[8] presented a broad spectrum of definitions of venture capital investment:

- investment in any high-risk financial venture;
- investment in unproven ideas, products, or start-up situations; i.e. the provision of what is called 'seed capital';
- investment in going concerns that are unable to raise funds from conventional public or commercial sources,
- investment in large and - in some cases - controlling interests in publicly traded companies where there is a considerable degree of uncertainty.

Liles also describes a dialectical phenomenon: "Interestingly enough, seed capital situations are considered by some individuals or firms as too risky to be described as suitable for venture capital and by others as the only form of a 'pure' venture capital investment opportunity."[9]

What can be noted is a steady trend within the venture capital industry to cover segments other than the traditional market: "The original concept of many prominent venture capital firms was that profits were to be made from financing

---

[8] Cf. Liles 1974

[9] Cf. Liles 1974, p. 492

men with ideas and helping start new businesses. As these investment firms gained experience, their interests have moved away from this type of venture."[10]

Today a comprehensive description of venture capital companies has to include the following activities:[11]

- investment in the seed, start-up and other early stages,
- investment in established companies that are unable to finance their expansion through banks or the stock exchange,
- investment in management buyouts and leveraged buyouts, and
- investments on the stock exchange where patient, supportive investment can facilitate ongoing business development.

Venture capital firms play three roles. The first stage is the identification of deals. For each 100 business plans that a typical venture capitalist might review, he will invest in only one. An experienced venture capitalist concentrates not only on the skills and ideas of the entrepreneur but also on his personal qualities. The second stage is the actual structuring of the deal and gaining an oversight of the firm. An agreement is reached which contains contractual elements including the type of financing (e.g., convertible preferred stock, common equity, etc.), the timing of capital infusions, explicit and implicit options, board representation and the advice provided. In addition to his formal role, the venture capitalist will often serve as an informal counselor to the management. The final stage is the harvesting of investments. During this phase, the investment is made liquid by performing an initial public offering (IPO), negotiating a merger, acquisition, buying back the entrepreneurial firm or liquidating the assets.

As a result of, and in addition to, providing capital, VC companies have another important function: providing management support to NTBFs. This is due to the fact that, in contrast to their competence in their fields of technology, the

---

[10] Loc. cit.

[11] Cf. Bygrave/Timmons 1992, p.1 ff.; Fetzer 1990, p. 5 ff.

management qualifications and experience of founders are generally relatively limited when a firm is in the initial phase of its development.

Services provided by venture capital companies may, for example, relate to the elaboration of marketing strategies and/or establishing contacts to external suppliers of resources and to clients. In extreme cases it may include all areas of management.

2.2 New Technology Based Firms

According to a broad variety of definitions[12] NTBFs can be defined as business start-ups, i.e. businesses not more than five years old with a yearly R&D expenditure of at least DM 100,000. However, absolute numbers do not provide a clear guideline for differentiation in all cases when the formation and the development of NTBFs is taken into account. Here the motivation behind, and the realization of, entrepreneurial activities in high-technology areas play a highly characteristic role in an NTBF.[13] Furthermore, the development process of an NTBF often varies considerably from that of an 'ordinary' company, for example, in the areas of sales, profit returns on capital and employment.[14] Due to the comparative character of this investigation, an NTBF is always considered in relation to quantitative and qualitative indicators.

Empirical analyses[15] show that the development process of NTBFs can be divided into five phases:
- the founding phase
- the R&D phase
- the market introduction phase

---

[12] Cf. Bachelier 1993; Gerybadze 1988; Kulicke 1987, u.a. 1993

[13] Cf. Kulicke u.a. 1993

[14] Cf. Pfirrmann 1994

[15] Cf. Churchill/Lewis 1983; Gerybadze 1988, 1991; Horvath et al. 1984; Kulicke 1987, u.a. 1993; Roberts 1991a

- the growth phase
- the stabilization and maturity phase

Each of these phases contains a specific emphasis on financing sources.

The development process of an NTBF begins with its initiators' intention to develop a high-technology enterprise which is economically viable and capable of surviving in the long term. Before this aim is achieved, the enterprise passes through the following succession of typical, ideal development phases, whose duration and characteristics may, however, vary greatly in practice:

The **founding phase** is a period preceding formal foundation. During this phase basic decisions are made about, for instance, the products offered and the markets of the potential firm. The need for capital at this stage is relatively small, so that it can usually be covered by the founders' personal savings or assets.[16] Unless the idea behind developing the business is discarded, the preparations for founding a firm are followed by its formal foundation.

The **R&D phase**, i.e. during which the innovative product is developed until it is ready for the market. The length of this phase primarily depends on the type of innovation and the NTBF's resources. However, it is generally very time and cost intensive.[17] There is, therefore, a great need for financing, which may run into several million DM. Due to the high degree of innovation risk, these capital requirements should be largely covered by equity and promotion funding. Thus, venture capital is an important source of capital funding.[18] Nevertheless, many VC companies in Germany are very cautious about making such investments at this stage. The main reasons for this are the high risks involved and the long time that

---

[16] Cf. Roberts 1991a

[17] Cf. Bräunling et al. 1993

[18] Cf. Roberts 1991b; Dunkelberg/Cooper 1983

still has to elapse before the NTBF will begin to make a profit.[19] Financing difficulties resulting from this situation may mean that NTBFs are not able to carry out development work. They may also lead to a loss of the competitive advantages that had been gained in the race against time.

The **market introduction phase** is a particularly critical stage in the development of NTBFs, since it is with market entry that the economic viability of the product, and thus also the firm's prospects of success become apparent.[20] Due to the length of time sometimes needed by potential customers to make decisions, and the possible need for a pilot phase, this stage lasts considerably longer in NTBFs than in other types of newly-founded enterprises. Although from the point of market introduction onwards some income accrues due to turnover activity in the firm, the capacity for self-financing is generally insufficient at this stage to finance the expansion of production and the sales organization. This again implies the need for large amounts of capital, far exceeding the requirements in the R&D phase, and may lead to considerable financing difficulties. Venture capital can, therefore, also be very important in meeting these capital needs. From market introduction onwards, however, NTBFs also increasingly finance themselves with debt capital.

The **growth phase**, i.e. in which other groups of clients or market segments are reached. This phase is characterized by the intensified development and expansion of individual functions, particularly marketing, manufacturing and controlling. This requires a further financial inflow in the form of debt capital and venture capital. Provided that business development is positive, the difficulties facing NTBFs' in procuring capital now decrease, especially as the risk for capital support has now become calculable.

---

[19] Cf. Horvath et al. 1984

[20] Cf. Wupperfeld 1993

The **stabilization and maturity phase** begins some years after its foundation. Every NTBF passes through a consolidation phase which does not, however, necessarily lead to a decrease in the growth of turnover. Rather, the organizational process stabilizes, as do the relations to customers, suppliers of goods and suppliers of capital.[21] With this stage of development the "expansion" phase comes to an end, so that the enterprise can no longer be described as an NTBF, but as an established high-technology firm. Venture capital may be required, for example, for making large scale investments, for re-structuring or for the first time a firm goes public.

To summarize, it can be said that

- NTBFs perform complex innovation projects, thus incurring high innovation risks (technical realization, marketing) and business risks (failure of the enterprise),
- NTBFs have a high demand for capital which has to be available in the long term in a form that maintains liquidity,
- it is difficult for investors, especially credit institutes, to estimate the volume of capital required, as well as the risks and the future development of the firm. This, together with the absence of real assets, means that banks are hesitant to finance the initial development phases of NTBFs,
- the injection of capital needed to deal with financing problems cannot really be reduced or spread over a longer time, since there would then be a danger that innovations could not be exploited, or that the comparative advantages gained from being ahead of competitors would be lost.

---

[21] Cf. Kulicke u.a. 1993

2.3   The Capital Market Segments in the US and Germany

In general, a variety of means for external financing exists in both countries. Different market segments can said to exist in the organization of the respective formal capital markets. The **leading markets** in the US are the registered exchanges such as the American Stock Exchange (AMEX) and the New York Stock Exchange (NYSE) which are by far the largest in the country. There are important regional exchanges in Chicago, Boston, Philadelphia, Cincinnati, Denver and on the Pacific Coast. The regional exchanges trade both stocks listed on the NYSE and local issues. All registered exchanges use the open outcry system of floor trading with principals and agents. For example the NYSE has 22 dual-capacity specialists who act as market-makers in about 100 stocks each.[22]

In Germany, the **first market** consists of the 'amtlicher Markt' (official market), where the prices are determined daily by the official brokers. There are eight regional stock exchanges: in Frankfurt/Main, Düsseldorf, Stuttgart, Hamburg, Bremen, Berlin, Munich and Hanover. Frankfurt is the most important of these, accounting for 71 percent, by value, of the turnover share on the German exchanges in 1994, having risen from 53 percent in 1987.

The **second market** in the US must be divided into smaller segments. The NASDAQ (National Association of Securities Dealers Automated Quotation System) is referred to in US parlance as the OTC (Over The Counter) market. The NASDAQ came into being in 1971 with the automation, constituting the nation-wide screen-based system, of the so-called Pink Sheet telephone market in small company stocks (which, in turn, grew out of local OTC markets). It now has one informal and two formal segments.[23]

---

[22]   Membership of the NYSE is on individual basis, but there are member firms which sponsor individual `seats´ of which there are 1,366 (Bannock & Partners 1994, p.85).

[23]   Cf. Bannock & Partners 1994, p. 86 ff.

The first segment of the **second market** consists of National Market System (NMS) stocks, which are regulated by standards similar to those on the NYSE and other registered exchanges. The second segment consists of Small Cap (or non-NMS) stocks, for which profit and turnover admission standards are lower, but which meet the same requirements as the NMS companies. The third informal segment consists of unregulated trading in small company stocks, for which the end-of-the-day trading prices are shown on NASDAQ screens. However, prices and the execution of deals are negotiated by telephone only. Trading on this third segment, which is the OTC in this terminology, is not included in any of the statistics available for the NASDAQ. The NASDAQ is a screen-based, quote-driven system providing for the automated execution for small orders and other on-line negotiation and execution services for orders of any size (SelectNet, ACES and ACT). There are some 470 active registered NASDAQ market-makers and 4,000 NASDAQ members trading through the system, as well as over 205,000 terminals leased from market data vendors which provide the best bids and demanded prices, latest trade details and index numbers. Apart from its central computer facility in Trumbull, Connecticut, and its back-up system in Washington D.C., the NASDAQ has no central location for trading, and its market-makers are dispersed throughout the US.

The **second market** in Germany consists of the 'geregelter Markt' (regulated market) where trading is similar to the official market. The prices are not determined officially but by independent brokers, though the official brokers also participate on some of the regional exchanges. Its introduction in 1987 was partly an attempt to improve the reputation of the so-called second tier market by means of stricter regulation.

The **third market** in the US, which dates from the 1950s, is an unregulated off-exchange market for block-trading between financial institutions and professional equity dealers via intermediaries.

The **third market** in Germany is the 'Freiverkehr' (free market or off-board trading). The free market has, as its name implies, a minimal regulatory system. Companies which apply do not need to show minimum size requirements or disclosure rules over and above those prevailing in Germany. The domestic shares quoted on the free market are considerably outweighed by dual listings of companies quoted on the official or regulated markets and on other exchanges, or by foreign shares.

The **fourth market** in the US is similar to the third market in that it facilitates large-scale block-trading between financial institutions and other professional equity traders, but differs in that the third market is restricted to non-exchange members. In the fourth market, which has grown rapidly since its foundation in 1969, traders are brought together by computer networks. Traders operate anonymously (hence `upstairs` for NYSE members). However, the block-trading of listed stocks between NYSE members is included in the NYSE turnover. The computer network caters for both listed and OTC stocks and combines both an order-driven system without intermediaries (crossing network) and market-makers' quotes for OTC stocks.

There is also a telephone exchange in Germany, which handles trade between banks and brokers outside the aegis of the market. It includes pre- and post-market trade in listed shares, domestic unlisted shares and some foreign shares. Most of the pre- and post-market trade disappeared with the introduction of a computer-based system, called IBIS, running parallel to the floor-based system. IBIS permits all-day trading (8.30 a.m.-5 p.m.). The bulk of trade is therefore in unlisted companies, of which there are about fifty.

Starting from the first segment, one can say that the formal restrictions on admissions (e.g., high levels of capitalization or extended records of profitability) for small companies and business start-ups are decreasing. Table 1 provides an overview of the basic requirements for the admission of securities to the US and German capital markets.

Table 1	Basic Requirements for the Admission of Securities to the US Capital Markets			
			NASDAQ	
	NYSE	AMEX	NMS	Small Cap
Market value of publicly held shares [1]	$18 million	$3 million	$3 million	$1 million
Minimum number of shares publicly held	1.1 million	0.5 million	0.5 million	100,000
Minimum number of shareholders	2,000 [4]	800 [4]	400-800 [3]	-
Minimum pre-tax income	$2.5 million	$0.75 million	$0.75 million	-
Trading history	3 years	3 years	- [4]	-
Number of market makers	NA	NA	2	2

Source: *Bannock & Partners*, 1994 and our own investigation

Notes: [1] Public float is defined by NASDAQ as shares not "held directly by any officer or director of the issuer or by any person who is the beneficial owner of more than 10 percent of the total shares outstanding".

[2] With 100 or more shares.

[3] Depending on the number of shares publicly held.

[4] Stocks of companies without current or even recent profits may be traded as NASDAQ NMS stocks if they have minimum trading history of 3 years and have net tangible assets of $12 million or more.

There are also NASDAQ initial requirements for net tangible assets (a minimum of $4 million) minimum share price at listing ($3.00) and other requirements for NMS inclusion.

Table 2	Basic Requirements for the Admission of Securities to the German Capital Markets		

	Official Market	**Regulated Market**	**Free Market**
Minimum percentage of equity to be held by external investors	The application must include all the securities of the same category. Also, 25 percent of the equity capital must be 'widespread', so as to ensure marketability. This latter requirement may be waived for large issues.	None	None
Minimum market capitalization required	DM 2.5 million	DM 500,000	None
Trading record	3 years[*]	1 year[*]	None
Continuing obligations over and above publishing annual accounts	An interim report covering the first six months of the financial year	None	None

Source: *Bannock & Partners*, 1994 and our own investigation

[*] Annual financial statements

In the US there is also a large informal market for (risk) capital. This, market operates through so-called business angels (wealthy private persons). However, no comparable information is available for the informal German capital market.[24]

---

[24] Cf. BVK 1994

3 Framework Conditions for Venture Capital: Results of the Literature Review

3.1 Development of the Venture Capital Markets - a Historical Overview

United States

Venture capital funds raise capital to invest in new business projects. These funds act as agents between the entrepreneurs who face search costs in locating funding, and uninformed institutional and individual investors. While venture capital comprises a relatively small percentage of capital market activities in the US, it provides an important source of funding for small businesses and offers the potential for high returns for investors. The industry has been responsible for helping to establish numerous successful enterprises. Among them are Apple Computer, Intel, Federal Express, Microsoft, and Lotus Development. Venture capitalists, unlike many other equity market participants, take active roles within their portfolio firms. In addition to the deal's origination, screening, evaluation, and structuring, they are responsible for monitoring the venture's post-investment activities on behalf of the investors in their managed funds. A venture capitalist often takes some form of a non executive managerial position within the portfolio company.

One can trace the beginnings of the US venture capital industry to the 1920s and 1930s, when wealthy families and individuals directly provided large sums of start-up money for companies such as Eastern Airlines and Xerox. The first organized venture capital firm was not founded until 1946, when Ralph E. Flanders, the president of the Federal Reserve Bank of Boston, and General Georges Doriot, a professor at the Harvard Business School, established American Research and Development (ARD) for the specific purpose of providing risk capital for new ventures.[25] In 1957, the firm invested $70,000 in exchange for 77% of common stock in a new company formed by four M.I.T. graduate students. By 1971, that investment had grown to comprise $ 355 million in common stock in Digital Equipment Corporation (DEC), which remains a global competitor in the computer industry.

In 1958 the Small Business Administration introduced the Small Business Investment Company, or SBIC program, as part of an overall effort to encourage economic growth through new company formation. The SBIC Act allowed SBICs to borrow four dollars at Treasury interest rates from the SBA for each dollar of equity capital they raised. While the SBA regulated all SBIC funds, the government was exempt from any involvement in investment decisions.[26] By 1965, the 700 licensed SBICs dominated the domestic supply of venture capital. However, incompetence and fraud plagued the industry, resulting in new regulations for the SBICs. By 1968, their number was reduced to 250, and accounted for only 21% of the venture capital pool. Private venture capital funds soon surpassed the SBICs in number and in the amount of capital they supplied. Today SBICs constitute only about 5% of the total capital pool.

The venture capital industry's growth was hampered by the recession set off by oil crises of the seventies. From 1978 to 1982, venture capitalists, entrepreneurs, and government joined in a combined effort to help revive the industry. Through legislation, they hoped to create a more favorable climate for venture capital markets. The 1978 Revenue Act lowered the capital gains tax rate from 49.5% to

---

[25] Cf. Bygrave/Timmons 1992

[26] Cf. Bygrave/Timmons 1992

28%, the first tax incentive for long-term equity investments since the late 1960s. The rate was later lowered to 20% by the 1981 Economic Recovery Tax Act.

Probably more importantly, three other acts removed many of the regulations governing the investment process. The 1978 ERISA "Prudent Man" Rule in effect allowed pension fund managers to invest in venture capital pools and other higher risk investments released a key source of new finance. Two additional laws were passed in 1980: the Small Business Investment Act and the ERISA "Safe Harbour" Regulation. The Small Business Investment Act reduced the reporting requirements for venture capital firms by redefining them as business development companies as opposed to investment advisers. The "Safe Harbour" Regulation dramatically reduced the risk exposure of venture capitalists by legally defining pension funds as limited partners, meaning that venture capital fund managers would not be considered fiduciaries of pension fund assets invested in the venture capital pools that they managed.

The question as to whether these acts were directly responsible for the subsequent growth of venture capital continues to be debated. Bygrave and Timmons state that "singularly and collectively, these five pieces of legislation completely revamped the venture capital industry, both immediately and throughout the next decade to the present...".[27] Others argue that the technological revolution which began in the late 1970s was responsible for the industry's revival. They both can be correct. The breakthroughs in microprocessing and biotechnologies, in particular, provided investment opportunities in a number of rapidly growing markets, from cellular communications to medicine. Even if this is so, however, there is no doubt that the 1978 Prudent Man Law in particular proved to be a defining piece of legislation. Pension funds have come to dominate the supply of finance to the venture capital market. Pension funds made up 15% of the capital committed to venture funds in 1978, and 46% in 1994. The large sums of capital from these institutions accounts for the tremendous growth by the venture capital industry in the 1980s. The rate of

---

[27]   Cf. Bygrave/Timmons 1992, p. 25

pension fund commitments - and the venture capital industry's growth rate - leveled out in the years 1990 to 1991 but has rebounded in the past few years (see Table 3).

Table 3	Sources of Capital Commitments (% of Annual Total) supplied to Private Independent Funds in the US						
	Total Capital* ($ billions)	Corpo- rations	Indivi- duals and Families	Pension Funds	Foreign	Endow- ments and Foun- dations	Banks and Insurance Com- panies
1980	0.661	18	17	29	8	15	13
1981	0.867	17	23	23	10	12	15
1982	1.423	12	21	33	13	7	14
1983	3.408	12	21	31	16	8	12
1984	3.185	14	15	34	18	6	13
1985	2.327	12	13	33	23	8	11
1986	3.332	11	12	50	11	6	10
1987	4.184	11	12	39	13	10	15
1988	2.947	11	12	46	14	12	9
1989	2.399	20	6	36	13	12	13
1990	1.847	7	11	53	7	13	9
1991	1.271	4	12	42	12	24	6
1992	2.548	3	11	42	11	19	15
1993	2.545	8	7	59	4	11	11
1994	3.764	9	12	46	2	21	9
Source: Venture Capital Journal, various issues							

*Excludes funds of funds

Financial market deregulation has also allowed banks to become more active in venture capital. The 1980 Banking Act broke down the distinction between commercial and investment banking, allowing commercial institutions to expand into securities transactions. Because of their substantial asset base, commercial banks are likely to become formidable players in investment banking.

In less than a decade, venture capital formation in the US had become almost entirely dependent on institutional investors, both foreign and domestic. One of the consequences of institutional investors in the market was that the funds increased in size. A large institutional investor might invest $25 million in a new fund. Such a large capital commitment entitled the institution as a limited partner to a great amount of negotiating leverage over the fund managers, i.e. the venture capitalists. As funds grew larger, growing from an average of $18 million in 1979 to $68 million in 1993, the average size of deals and investments also increased. Consequently, funds had to invest relatively large amounts of capital. Venture capitalists began to move away from start-ups to later-stage financing that required more funds per deal (see Tables 4 and 5). The explosion in leveraged buyouts (LBO) during the eighties encouraged the trend toward bigger deals. In addition, the increased role of institutional investors may have led to a change in venture capitalists' time horizons. It can take at least five years or longer for a start-up to reach a harvest stage, while a more mature company may show a profit in a couple of years. Institutional investors often demanded earlier and more frequent cash distributions, lower fees for the venture capitalists, and often a minimum return from their investments before the venture capitalist could participate in the transaction. The dominant force of pension funds and other institutions in venture capital partnerships led to a fundamental restructuring of the business.

Table 4	Capital Raised by Venture Capital Partnerships and Investment Stages in the US			
	Total Capital* ($ billions)	Percent of Number of Partnerships by Investment Stage		
		Seed	Balanced	Later
1980	0.661	35	61	4
1981	0.867	43	57	0
1982	1.423	38	57	5
1983	3.408	32	59	9
1984	3.185	34	59	7
1985	2.327	37	49	14
1986	3.332	41	49	10
1987	4.184	32	60	8
1988	2.947	41	55	4
1989	2.399	50	45	5
1990	1.847	14	72	14
1991	1.271	48	47	5
1992	2.548	36	40	24
1993	2.545	22	66	12
1994	3.764	30	44	26

Source: Venture Capital Journal, various issues.

*Excludes funds of funds

Table 5	Investments by Stage (% of Annual Total) in the US						
	Disburse- ments ($ billions)	Seed	Start- Up	Other Early	Expan- sion	Bridge Loans and Public Purchases	LBO Acqui- sitions
1989	3.26	4	9	15	47	4	21
1990	1.92	3	7	17	52	3	18
1991	1.36	4	6	22	54	11	3
1992	2.54	3	8	13	55	14	7
1993	3.07	7	7	10	54	16	6
1994	2.74	4	15	18	45	12	6

Source: Venture Capital Journal, various issues

Germany

The German venture capital market came into being in the 1960s as a result of an ongoing discussion on the part of the banks over private capital deficits and sinking private capital quotas for medium-sized companies and savings banks. This led to the creation of the first stock cooperations[28] on this market to provide private capital or similar funds to medium-sized firms (i.e. firms with about 1,000 employees or an annual turnover of DM 100 million) that were not in a position to go public. Contrary to expectations, there was practically no investment in small and infant firms, particularly new technology based firms (NTBFs).

---

[28] For example: Deutsche Beteiligungsgesellschaft m.b.H. (DBG), 1965, Allgemeine Kapitalunion GmbH & Co. KG (AKU), 1966, KBG Kapitalbeteiligungsgesellschaft m.b.H., 1968, Gesellschaft für Beteiligungen und Kapitalverwaltung m.b.H. & Co., (GeBeKa), 1969, Beteiligungsgesellschaft für die deutsche Wirtschaft m.b.H., 1969. Among the savings banks 11 Kapitalbeteiligungsgesellschaften (KBG) were founded between 1966 and 1976.

A second phase in the development of the German venture market began with a founding wave that occurred in the early seventies. Since 1970, the German Ministry of Economic Affairs, as the caretaker of ERP[29] special funds, has been offering refinancing funds at favorable interest rates as well as return guaranties for non-profit guarantee societies if they become involved in medium-sized firms. The expansion of the ERP program aimed at making it possible to use factors of economic promotion to prompt the funding of capital joint venture companies. This was intended to help secure existing medium-sized firms, as private societies did not use the ERP special funds offered to the extent anticipated. The Federal Laender prompted the creation of Mittelständische Beteiligungsgesellschaften (MBG), i.e. investment companies that aimed to provide equity capital to SMEs, especially in the form of silent partnerships. MBGs initiated business activities in all Federal Laender in the seventies and eighties. Right up to the present, the most active MBGs have been in Baden-Württemberg, Hesse and Bavaria, whereas the MBGs in the other Federal Laender had already either limited or even terminated their investment participation in the second half of the seventies.[30]

However, due to the ERP guidelines, the MBGs did not invest in NTBFs. With the result that these firms did not dispose of any venture capital. In particular, there was a demand for risk capital that was geared towards long-term value growth and thus did not have such a high liquidity-requirement during the first three years of development of portfolio firms.

To respond to this demand adequately and to gain experience with early stage investments in NTBFs, the Deutsche Wagnisfinanzierungsgesellschaft (German Venture Investment Company, WFG) was established in 1975.[31] The WFG can be regarded as the origin of the German Venture Capital industry. Twenty nine

---

[29] European Recovery Program

[30] Cf. Mayer/Müller 1991

[31] See for an intensive discussion of the WFG: Mayer/Müller 1991.

German banks participated in the foundation contract of the WFG and the Federal Government. The latter committed itself to balancing possible losses of the WFG and thus minimized the risks for private capital. The main aim of the WFG was to provide venture capital for inventions and technological developments in small and medium sized firms. In addition to this the portfolio companies were to receive management assistance and network support, i.e. contacts to suppliers, customers and relevant services. For a number of reasons, which included - for instance - problems related to the original fund construction and management conception, the WFG changed its structure and strategy in the mid-eighties. In the mid-nineties the WFG is now only supervisor of current investments.

The third phase in the development of the German venture capital market started in the early eighties. It came about as a result of information and discussions over what the successful American venture-capital model had to offer. In the case of private capital and management that were supporting young high-technology firms with a high growth potential, the successful examples of Apple or Genentech (including the figures for turnover, profit and participation effects of firms financed by venture capital) also triggered venture-capital euphoria in Germany. 1983 was the year in which the German venture capital scene was born, when associated companies such as IVCP and TVM[32] were created by banks and industrial firms following the US model. Furthermore, by the beginning of the eighties the classical German investment companies were vigorously participating in the early phase of high-technology firms. Foreign venture capital firms also appeared on the German market in the second half of the eighties.

With this third phase, what may be considered as the rapid development phase of the German venture capital market also got under way. Starting around 1964, this continued of twenty years, as subsequent calculations were to show, resulting in a

---

[32] International Venture Capital Partners S.A. Holding, Luxembourg or Techno Venture Management, Munich, respectively.

portfolio of up to DM 1 billion. Four years later this figure had reached DM 2 billion and rose to more than DM 5,3 billion (funds under management) in 1994. From 1983 to 1994 the market volume increased by a factor of seven; a factor which includes the number of partnerships, which, in turn, increased by a factor of 2.6 to 2.800. The rate of increase was especially high during the last six years (see Table 6).[33]

Table 6	Long-Term Development of the German Venture Capital Market											
Year	1983	1984	1985	1986	1987	1988	1989	1990	1991	1992	1993	1994
*)	785	867	1156	1360	1592	1974	2577	3221	4029	4553	4982	5342
**)	1069	1138	1288	1429	1583	1536	1752	1977	2298	2455	2657	2780
Source: BVK (ed. 1994)												

*) Funds under management (in DM million)

**) Number of partnerships

Despite the great number of investments, not only in the portfolios of classical investment companies, but also in those of the new venture capital companies, the investments in NTBFs proved to be both unsuccessful and disappointing.[34] This was largely a result of the limited experience the investment managers had in the selection and management of NTBFs. What had began as euphoria quickly faded and a mood of sobriety set in. Some of the venture capital societies that had been

---

33 Cf. BVK 1994

34 Cf. Schmidt 1988

created in the eighties had never really developed their full business potential; others ceased their activities after a few years.[35]

This high failure rate in relation to the effort put into financial participation and management support to NTBFs resulted in a relatively quick reorientation in investment politics. German investment companies concentrated on financing expansion, participations in established firms in traditional economic fields and management buyouts (MBO), i.e. later-stage development finance. Many companies reduced or stopped their participations in NTBFs completely. As a result, these firms were no longer supplying any risk-sustaining capital worth mentioning after 1989. Only the MBGs in Baden-Württemberg and Hesse continued to finance NTBFs by adhering to the ERP refinancing program and by refinancing offers in the respective Federal Laender.

By changing the portfolio structure, venture capital companies sought to increase their turnover of investments in order to cover management costs. With the change in both target groups and the size of participations, the support strategies also changed, i.e. the size and intensity of the portfolio firms declined ("hands-off" instead of "hands-on"). Despite the differences and investment targets that existed, this resulted in an equalization of the investment policies of most of the German investment companies and also evened out the differences between VC companies and the other forms of investment companies.[36]

The last few years have, however, once again seen a development in the seed-capital market. The pilot scheme "Beteiligungskapital für junge Technologieunternehmen" (BJTU, share capital for young high-technology firms) has played a decisive role in this respect. This model has been helpful in the founding of some of the seed-capital companies. It has resulted in several existing joint venture companies returning to the scene and becoming involved once more in

---

[35] Cf. Bräunling, Gerybadze, Mayer 1989

[36] Cf. Frommann 1992

the early development phases of NTBFs.[37] Despite public backing, seed and start-up financing, taken together, have always represented a small share of total venture capital investments. In fact, according to Table 7, they have access to approximately 10 % of the total portfolio (in relation to investment volume), which was - and remains - relatively unimportant. In comparison their start-up in 1985 was still running at 25 %.[38] The lion's share of the value of the total German portfolio was - and still is - invested in expansion financing. Management buyouts and management buyins came second and were lagging ever further behind.[39]

Table 7   Investments* by Stage (% of Annual Total) in Germany

Financing Phase	1994	1993	1992	1991
Seed	2,14	1,20	1,10	2,26
Start-Up	7,99	7,22	5,55	3,47
Expansion	48,95	61,79	44,92	66,86
Bridge	2,82	2,79	10,57	1,26
MBO/MBI	35,28	24,66	22,12	13,32
Turn-Around	1,87	0,67	4,23	0,96
Sum	99,05	98,33	88,49	88,73
not labeled/others	0,95	1,67	11,51	11,87
Total Capital (DM million)	1998,46	1111,95	1229,17	995,49

* Gross investments, i.e. first-round and follow-up investments

Source: *BVK*, various issues

---

[37] Cf. Harnischfeger, Kulicke, Wupperfeld 1992

[38] Cf. Frommann 1993

[39] From the information received from Bundesverband Deutscher Kapitalbeteiligungsgesell-schaften e.V. (BVK, German Venture Capital Association) it has to be taken into consideration that there is a misunderstanding in the characterization of the volume of shares in the respective segments of the market especially in the middle of the eighties. Here the comparison of data is limited. The BVK allocates seed and start-up financing to non- technological young firms, which leads to a reduced share of financing for the young firm in the early development phases. With respect to the number of participations, seed- and start-up investments make up a very significant share with 31 % of the total.

In Germany, the investors consist of statutory private and public banks and are, in fact, responsible for more than half of the volume of all investments in firms made by investment companies (see Table 8). The insurance companies and industrial enterprises, which are skeptical about venture capital, as well as government institutions and other investors, only play a subsidiary role. In the past, and in the present too, a more attractive investment for these large capital sources has been seen in traditional investment funds, e.g., in real estate projects.[40] At a European level, credit institutes are the most important capital lenders for investment companies. However, they do not have the same central importance that they do in Germany. Despite an on going discussion about growing influence of pension funds in Germany they have not played an important role so far. Pension funds constitute the second most important type of investment in Europe.[41]

Venture capital is still going through its developmental phase in the new Federal Laender,. As a result, no data can be given for the main capital lenders. On one hand, West German and foreign investors have deposited very large amounts in firms, whilst on the other hand, MBGs are being established in the new Federal Laender. It was in 1990 that investments were first made in the new Federal Laender. In 1994 there was a portfolio of DM 560 million with 382 participants. In comparison to the old Federal Laender, however, the requirement and demand for joint venture capital is still very low there. Thus NTBFs are scarcely being financed in the new Federal Laender. The reason for the very low involvement in investment companies lies in the threefold issue of unsettled property rights, administrative obstacles and the generally very high costs of finding capital as well as the risks for the venture capital companies.[42]

---

[40] Cf. Workshop 1983; Wirtschaftswoche 1995

[41] Cf. EVCA 1993

[42] Cf. EVCA 1992

Table 8	Sources of Funds for Venture Capital by Sectors (% of Annual Total) in Germany			
Capital Provider	**1994**	**1993**	**1992**	**1991**
Private Banks	34,36	34,37	33,59	43,02
National Banks	16,65	13,76	14,92	17,85
Savings Banks	4,11	4,12	3,13	5,84
Insurances	11,71	12,12	11,52	9,57
Industry	7,84	9,27	5,61	5,75
Private	8,24	6,66	5,40	13,64
State	6,97	5,62	3,76	0,06
Other	3,36	3,39	1,46	4,27
Sum	93,24	89,29	79,39	100,00
no details available	6,76	10,71	26,61	
Total Capital (DM million)	**8828,59**	**8257,86**	**7855,48**	**5550,98**
Source: *BVK*, various issues				

3.2   The Markets Today

United States

Today there are around 600 venture capital firms in the United States, controlling over $30 billion in capital. One of the most dramatic changes which occurred during the 1980s was the globalization of capital markets. In 1979, venture capital was virtually non-existent outside the United States. A decade later, the amount of capital raised in 1983 by Europe including the UK had exceeded that raised by the venture funds of the United States ($4.1 billion and $2.95 billion, respectively).

The increase in the supply of venture capital funds and the number of firms, has meant a number of changes in the US industry, as described above. In general, there has been a shift in the relative bargaining power from the venture capitalist to both the entrepreneur and the institutional investor. The market has also become more heterogeneous. Prior to 1980, the several hundred firms shared similarities in the sizes of the funds, investments, sources of capital. Start-up and early stage investment has declined since 1985. In 1980, investments were targeted in the computer hardware industry, while today computer software and biotechnology have become the dominant sectors (see Table 9).

Table 9	US Venture Capital Industry Distribution by Dollars Invested 1994			
Industry	Total Invested (in $ millions)	%	Number of Companies	%
Biotechnology	303	11	101	10,0
Commercial communications	46	1,7	35	3,5
Computer hardware/software	171	6,2	72	7,1
Consumer-related	173	6,3	73	7,2
Energy-related	2	0,1	4	0,4
Industrial automation	16	0,6	6	0,6
Industrial products and machinery	56	2,0	40	4,0
Medical/health care-related	473	17,3	167	16,5
Other electronics	167	6,1	63	6,2
Other products and services	663	24,2	110	10,9
Software and services	378	13,8	225	22,3
Telephone and data communications	294	10,7	115	11,4
Total	2.741	100	1.011	100

Source: Venture Capital Journal, various issues

Bygrave and Timmons note that the mentality of the industry has also changed becoming more risk averse.[43] There appears to be less of a focus by the venture capitalist on the deals that offer spectacular returns. This is in part due to the recognition by fund managers that competition among funds for deals has made such opportunities rare. There is also a growing preference for investing in established ventures, which in general require less risk, offer quicker payoffs, and by extension, generate lower returns. Today's financing now involves a combination of venture capital, leveraged, and management buyout deals, with a continuing increase in the use of debt. Another result of the increase in the number of funds is that many firms specialize in particular investment stages, or in particular industries. The advantages of such specialization for fund managers include not only experience and knowledge, but also contacts with suppliers, customers, or engineers for an industry. This holds true especially for early stage investments. It is less relevant for later stage investment, for example in LBOs. With the ability to offer strategic guidance in addition to capital, the venture capital firm can add value to a project.

There are three main classes of venture capital organizations:
- independent funds,
- Small Business Investment Companies (SBICs), and
- corporate subsidiaries.

Table 10 shows the breakdown of the industry by type of fund structure.

---

[43] Cf. Bygrave/Timmons 1992, p.45

Table 10   Composition of the US Venture Capital Funding Pool

	Venture Capital Stock ($ billions)	Percent of VC Stock Managed by:			
		Independent Partnerships[a]	Corporate Financial[b]	Corporate Industrial[c]	Independent SBICs[d]
1980	4.5	40	31[e]		29
1981	5.8	44	28[e]		28
1982	7.6	58	25[e]		17
1983	12.1	68	21[e]		11
1984	16.3	75	12	9	4
1985	19.6	78	13	8	3
1986	24.1	81	11	8	3
1987	29	83	11	7	2
1988	31.1	83	14	7	1
1989	34.4	79	14	7	0
1990	35.9	80	13	7	NA
1991	32.9	81	12	7	NA
1992	31.1	93	2	5	NA
1993	34.8	81	7	12	NA
1994	34.1	79	14	7	NA

Source: Venture Capital Journal, various issues

[a] Independent partnerships also include a few incorporated venture capital firms.

[b] Corporate-Financial includes bank-affiliated SBICs and partnerships managed by affiliates of financial institutions. After 1992, investment bank-affiliated partnerships were classified as independent partnerships.

[c] Corporate-Industrial includes venture capital subsidiaries of industrial corporations, including affiliated SBICs.

[d] Independent SBICs are included in independent partnerships after 1989.

[e] Corporate financial and corporate industrial are reported together.

Independent Funds

Independent funds most often take the form of limited partnerships. The venture capitalist acts as a general partner, screening potential investments and guiding the management teams of the partnership's portfolio firms. The general partner organizes the fund and accepts full legal liability for its management. In addition, the venture capitalist will typically put up 1% of the funds raised. The limited partners are the investors, providing financial capital for the fund. They may be private individuals, pension funds, endowments, and insurance companies. The partnership has a finite life, usually 10 years, at which time the funds must be distributed to the limited partners. Sometimes the life may be extended for several years. The venture capitalist receives annual compensation in the form of 2 to 3% of the total committed capital pool and 15 to 25% of the carried interest (the realized capital gains) usually after a minimum investment performance level has been achieved.

Small Business Investment Companies (SBICs)

SBICs, as mentioned before, are licensed and regulated by the SBA. They are closed-end investment trusts that provide both financing and managerial assistance to start-up firms. Investment income from these trusts is not taxable until it is distributed to shareholders. Banks and insurance companies find SBICs an attractive investment vehicle due to this tax-deferrable aspect. It is uncommon for individuals to invest through SBICs.

Corporate Subsidiaries

For large corporations, banks and industrial companies, investing in start-up firms provides the opportunity for diversification, new technologies, markets and generated attractive capital gains.[44] Most corporations have participated in venture

---

[44]   Cf. Block/MacMillan 1993, p.350

capital markets through corporate subsidiaries, such as Exxon Enterprises and Gevenco (a subsidiary of General Electric). These two funds, like many corporate funds, have since been abolished as the corporate strategy has changed. Other companies invested directly in firms, which may have led to closer ties to the firms in their portfolio. In many cases, however, the corporate venture arms did not get offered them, or became embroiled in battles with corporate research officials, who desired that the resources be devoted to internal programmes.

The process of venture capital funding consists of multiple stages. Each successive stage is typically associated with a significant development in the company, such as the completion of a design, or the emergence of the firm into profitability. The venture capitalist essentially buys a call option on further investment, which gives the entrepreneur an incentive to perform. In the first round, or seed stage, the venture capitalist faces many risks associated with the unknown aspects of the project. This initial stage often precedes the formation of a complete management team, product design, or market testing. By the second round of financing (on average after the first nine months of operation), these risks will have been greatly reduced as the investors will have more knowledge about where the project stands on these aspects. For the venture capitalist, the trade-off of the lower risk is a higher price for shares, and therefore a lower expected rate of return. The third round of capital formation often occurs after a product has been produced. At this point, most of the market research has been completed, and therefore the risk is again reduced, resulting in a further increase in the investment's valuation.

Table 5 had shown the distribution of funds for these stages. Expansion financings have continued to dominate the US market. Staging capital investments in a project can be seen as an advantage for both parties. From the venture capitalist's point of view, he gains the option to revalue or abandon the project in the context of new information. While, from the perspective of the entrepreneur without staging his profits would have been greatly diluted by the venture capitalist who would have demanded a higher share of the company's gains, having faced only the very high risks of the first round. There is also the possibility that such risk simply would be prohibitive to many venture capitalists. Sahlman notes that "in the case where there is great uncertainty and the investor has the opportunity to stage capital investment over time, the value of the option to abandon might be sufficiently high to change the net present value from negative to positive".[45]

Exits/Disinvestments

IPOs and mergers and acquisitions (trade sales) are the two primary exit mechanisms for liquidating portfolio companies (see Table 11). The most common strategy to exit or "harvest" an investment is through an IPO, or initial public offering. Venture capitalists generate the highest returns from firms go public.[46] the company issues stock, thereby becoming a public enterprise. Other exit strategies include mergers, liquidations, or share repurchase. Because of the importance of IPOs in harvesting, the liquidity of venture capital investments depends a great deal on the stock markets, particularly the NASDAQ exchange. In the strong market of 1983, around 700 small companies issued IPOs, raising nearly $6 billion. The robust market of the early and mid-eighties led to strong rates of return on venture capital funds formed in the 1970s. However, the IPO binge ended with the stock market crash on October 19, 1987, and new issues in 1988 and 1989 fell by 75% from the level of 1983. With the decline of the IPO, alternative exit strategies developed. For the years between 1987 to 1989, acquisition exits outnumbered IPOs. In the late 1980s and 1990s foreign investors provided American venture capitalists with another option. Japanese companies, with the advantage of the

---

[45] Cf. Sahlman 1989, p. 29

[46] According to a Venture Economics study, a $1 investment in a firm that goes public will provide an average return of $1.95 in addition to the initial investment (with a holding period of 4.2 years). An investment in an acquired firm yields a cash return of 40 cents for a holding period of 3.7 years; see Lerner 1994b.

strong yen, sought access to technology. Major the US firms also became formidable players in the acquisition market. While by the spring of 1991, biotechnology stocks helped trigger an IPO market revival, larger firm purchases have continued to be a major source of exits for domestic venture capital.

Table 11	Number of Venture-Backed New Issues and Acquisitions in the US													
Year	80	81	82	83	84	85	86	87	88	89	90	91	92	93	94
IPOs	27	68	27	121	53	47	98	81	36	39	42	122	157	165	136
Acquisitions	28	32	40	49	86	101	120	140	135	136	88	65	69	57	97

Source: Venture Capital Journal, various issues

A study of venture-backed biotechnology firms between 1978 and 1992 found that exit strategies are highly dependent upon equity valuations.[47] Venture capitalists take firms public when equity valuations are at their highest, or turn to the alternative of private financings when valuations decline. More experienced venture capitalists seemed to have a better ability at timing IPOs. This may be due to their proficiency in recognizing when valuations peak. An alternative explanation might be that these seasoned venture capitalists simply have advantages over less experienced colleagues. They have important connections to investment banks, and less pressure to generate quick returns in order to attract investors for future funds, which affords them the time to wait until the market is optimal.

---

[47] Lerner's study was based on a sample of 350 privately held venture-backed biotechnology firms financed between January 1978 and September 1992; ibid.

Returns on Capital

Apple Computer is often cited as an example of the classic "super deal." During 1978 and 1979, several venture capitalists invested a total of just over $3.5 million in the start-up company. When Apple went public in 1980, the value of their investment was $271 million. While there have been similar dramatic stories of returns realized in other companies, such as DEC and Lotus, these are the exception. A venture capital firm generally relies on a few investments to offset losses. A 1988 study by Venture Economics found that for the 383 investments made by 13 firms between 1969 and 1985, more than one-third ended in an absolute loss. However, 6.8% of the investments resulted in payoffs greater than 10 times the cost and yielded almost 50 % of the ending value of the funds' portfolio.[48]

The guidelines for earning high rates of return are investing early, avoiding dilution, and picking companies with enormous upside potential. Ideally, a company would have some form of exclusive right to a specific market, e.g., through a patent or license as is sometimes the case in high-technology industries. Investors are also concerned with the entrepreneur and management team, as they are important factors in whether or not the venture will ultimately succeed. Sahlman notes that an important factor in the profitability of an investment is the time between the initial investment and the return on capital. A particularly important ability for the venture capitalist to be able is to determine whether or not to continue to invest in a struggling company. Federal Express is an example of a successful company which at one stage seemed doomed to failure.

Paul Gompers offers several explanations for the general decline in returns since the early 1980s.[49] The success of venture capital firms, an active IPO market, and financial deregulation led to a dramatic increase in the supply of capital from

---

[48] Cf. Sahlman 1990, p. 483

[49] Cf. Gompers 1994a

pension funds. This put pressure on venture capitalists to find deals, thereby increasing the valuation of these deals in a competitive market. In addition, the nature of pension funds required quicker returns and less risk, leading to relatively later-stage investments, and therefore lower expected returns.[50] According to one survey, one of the effects of the growth of the venture capital industry was a reduction in the quality of decision-making on the part of the venture capitalist. The increased competition for deals has forced these funds to spend less time on the screening and deal-making process, and has also led to a dilution of returns, with the increase in the price demanded by entrepreneurs.[51] Table 12 shows the average rate of return realized by limited partners in recent years.

Table 12	Annual Returns: One-Year Net IRRs for Mature Funds formed 1969-1990 in the US			
Year	1991	1992	1993	1994
IRR (%)	24	12,5	19,7	16,2
Source: Venture Capital Journal, various issues				

Germany

Today there are different types of investment companies on the German venture capital market (see Table 13). A classification, mainly based on initiators and investors, commercial criteria (dividends and economic promotion), and financing phases reveals the following types of investment companies:[52]

---

[50] Cf. Barry 1990, p. 6

[51] Cf. Premus 1984, p. 14

[52] This classification specifies the German venture capital market on a very detailed level. Another less detailed classification would lead to at least three different types of investment companies (for example Business Investment Companies, MBGs, VCCs).

Table 13: Types of German Investment Companies

Charac-teristics	Types of investment companies	Business investment companies of banks and insurances	Business investment companies of the savings banks	MBGs	Venture-capital companies	Seed capital companies
Initiator/Investor		Banks and insurance companies	Savings banks and Union banks	Federal governments, other public insurances, credit institutes	Credit Institutes, Industry, Share Managers	Share Manager, Industry, Credit institute
Aim of share		Profit orientated (continued turnover)	Profit orientated (continued turnover) influenced also by economic support	Economic support	Profit orientated (value increase)	Profit orientated (value increase)
Main focus of investment		Firms with growth potential	Firms with growth potential	SME's	(innovative) firms with a growth potential	Seed- and Start-up-Phases of JTUs
JTU significance		None	Limited	Limited	Limited	Very important
Support service		Little	Little	Very little	Complete	Complete
Forms of shares		Silent and direct	Silent, partly direct	Silent	Direct	Direct and silent
Time period of creation		In the 60's, with the greatest intensity in the 80's	In the 60's, with the greatest intensity in 80's	70's and 80's in the NBL after 1992	From the beginning to the middle of the 80's	At the end of the 80's
Geographical orientation		National, no main focus	Area-covered by the savings banks or their headquarters	The respective Federal Land	National, partly international	Partly regional, partly national, a few also international

Source: FhG-ISI

Business Investment Companies Established by Banks and Insurance Companies

The industrial investment companies are independent stock corporations created by banks and insurance companies. The industrial investment companies established by the banks have as their main goal an increase in the return on investments for all shares. They also partly have strategic goals such as acquisitions and improving client relations (portfolio firms as potential creditors). Since the early eighties, factors such as gaining experience in the analysis of innovative firm concepts for credit financing and creating better company images have also played a role.[53] The prime investment goal of the banks' industrial investment companies comprises the financing of established firms with growth potential. In practice, they did not invest in young and small firms, whose capital requirement is relatively low. The reasons for this are the immense difficulties involved in calculating the risks involved as well as the unfavorable management cost ratios with respect to the possible choices, supervision and share dividends. The banks' industrial investment companies only provide a limited amount of business consulting for these portfolio firms. The insurance companies also participate in the funding of industrial investment companies of their own subsidiaries[54] within the scope of their own limited possibilities.[55] Owing to the limited synergy effects between the investments performed by industrial investment companies and the insurance companies, the holding companies established by the insurance companies aimed almost exclusively at the highest possible dividents. Otherwise their share policies are very similar to the investment participations of the banks.[56]

---

[53] Cf. Büschgen 1985; Grisebach 1989

[54] For example Hannover Finanz GmbH (1979, a subsidiary of the Equity Investment Company of the German Industry), Kapitalbeteiligungsgesellschaft der Deutschen Versicherungswirtschaft (KDV).

[55] Insurances are governed by the capital investment regulations of the Insurances Governance Laws (VAG). According to the § 54 VAG 1987 GmbHs and OKGs as well as silent partners are allowed only to hold a maximum of up to 5 % of the deposited capital; see Grisebach 1989.

[56] Cf. Kulicke u.a. 1993

The Business Investment Companies Established by the Savings Banks

The business investment companies established by the savings banksprimarily aim at acquisitive economic goals and targeting regional economic support.[57]   In the latter case, they at least offer silent partnerships, i.e. investments on favorable terms to small and medium-sized firms in their respective regions. Here they generally apply the refinancing offers from the ERP funds. At the beginning of the eighties, the business investment companies of the savings banks showed a high degree of involvement in the financing of NTBFs in contrast to the style of founding companies and technological centers that was prevalent at the time. Today they invest in much the same way as the business investment companies of the banks and insurance companies, i.e. they mostly invest in companies with growth potential. Only a few of the original participations still have substantial shares in NTBFs.

MBGs[58]

MBG funds are business investment companies whose economic policy aim is to distribute silent partnerships, i.e. investments to small and medium-sized firms. The MBGs mainly receive support - as economic self-help institutions - from industry, the chambers of commerce, the provincial banks and the regional credit institutes. They work in close cooperation with the association of credit institutes. They do not have financial funds of their own worthy of note, but largely refinance themselves through support programs provided by the state and the Laender, however, the most important source of funds are those of the ERP. As a result, the investment conditions with respect to share volumes, the life-expectancy of firms, forms of investment, dividend procedures and agreed repayment schedules are largely determined by the framework in which the ERP programs are carried out. In case of

---

[57]   Kapitalbeteiligungsgesellschaften (KBG)

[58]   See for a first explanation, p. 28

default, the firms receiving aid can fall back on the provincial guaranties through the security banks or guaranty societies.[59] [60]

As a result, the prime focus of investment for most MBGs lies in the financing of medium-sized firms. The MBGs in Baden-Württemberg, Hesse and Bavaria as well as the innovation fund Berlin also finance, within the framework of the BJTU pilot scheme and the respective regional support program, the early development phases of NTBFs.

Venture Capital Companies

Venture capital companies (VCC) - as opposed to the business investment companies of the banks, the savings banks and insurance companies - are independent leading investors that distinguish themselves from KBGs and MBGs with respect to their managers as well as their investors. These investors are generally industrial firms and credit institutes. In the last few years, foreign VCCs have increasingly founded subsidiaries in Germany. After having primarily financed NTBFs in the early eighties, VCCs investments in NTBFs now only play a secondary role. The main focus of their investments lies in the financing of the growth phases of medium-sized (as well as technologically oriented) firms, not to mention management buyouts and turn-arounds. As a result, the relation between the investment volume and the effort required for valuing and management is more favorable with regard to portfolio firms than it is in the case of NTBF.

VCCs have a high dividend orientation and participate directly in investments where profits are only realized through pay-offs to a small degree, yet largely realized as capital gains through the sale of shares. Hence, due to their international

---

[59] For the development and the working procedures of the security banks; see Schütt 1993.

[60] Amounts to 70 % of the total share. BJTU supported shares under the framework of the pilot scheme, 90 % of the risks of the MBGs are taken over by the Kreditanstalt für Wiederaufbau (KfW).

networks and their interdisciplinary, pooled management teams, they can advertise that they are in a good position to give their portfolio firms comprehensive management support. As this kind of investment activity first picked up on the German market in the eighties, they do not have a long tradition of accretion investments there.

Seed Capital Companies

Seed Capital Companies (SCC) are a special form of VCCs. However, they distinguish themselves from the latter by primarily focusing on investing in the early development phases of NTBFs. These relatively young funds are generally both independent and based on initiative coming from two sources: managers and credit institutes. The managers generally come from a field of activity very similar to that of the high-technology companies and are prompted by public support programs for seed capital participations[61], to undertake new investments. The other prime source of initiative comes from the credit institutes.[62]

SCCs aim at a high returns which are realized through capital gains. In addition to this, some also target economic support programs. Seed funds generally finance their portfolio firms directly or, in part, through silent investments and offer these firms active and comprehensive management support. By pursuing investment policies of this nature, their activities come closest to the classical venture capital concept when compared with all the different types of industrial investment companies. Due to the fact that NTBFs can only find capital lenders with great difficulty in their early development phase, the SCCs also have to assume the

---

[61] Primarily the pilot scheme share capital for young technological firms BJTU and the European Pilot Scheme for the stimulation of seed capital. The "European Pilot Scheme for the Stimulation of Seed Capital" that was started by the EU General Directorate XIII at the end of 1988 supported the creation of seed capital cooperations through repayable subsidies of up to fifty percent of the firms requirements/management costs, these subsidies were limited in time

[62] Cf. Kulicke u.a. 1993

function of providing initial financing. In such cases they act as the seed funds' "intelligence", i.e. private capital coupled with intensive consultancy support.

Table 14	Exit Channels in Germany for Venture Capital Funds (% of Annual Total)		
Exit*	1994	1993	1992
buy back	34,37	39,14	35,3
trade sales	39,37	27,96	27,94
secondary purchase	2,53	8,86	8,82
going public	11,96	16,86	11,41
others	11,77	7,18	16,53
Total Capital (DM million)	**530,13**	**402,71**	**251,61**
Source: *BVK*, various issues			

* Data from 39 participants for 1992

  Data from 41 participants for 1993

  Data from 40 participants for 1994

Exits/Disinvestments

The most common methods of disinvesting in Germany are repurchasing shares and performing trade sales. Table 14 shows that IPOs accounted for a relatively high share in 1993 (17 %) if one takes the other years and the two primary exit mechanisms - buying back shares and trade sales - into consideration. Nevertheless, when compared to the absolute number - eight in 1992, ten in 1993 and twelve in 1994 - it becomes clear that this method of going public does not play such an important role in the case of VC disinvestments. On the other hand, if one proceeds from different valuation criteria for the US and German VC managers, it becomes apparent that rather conservative market expectations often hamper the activities of ·

potential shareholders in Germany. Trade sales became less important during the last recession but increased again in 1994. While most large companies were forced to focus on "core competence" centers, diversification strategies - including the acquisition of NTBFs - decreased for a short period. A considerable loss in 1994 is ascertainable for secondary purchases. This underlines the minor and continually decreasing role of this method of exiting from an investment.

Table 15 shows the investment strategy of German VC funds. For example, the mechanical engineering/machine tool industry and the trade sector have the biggest share of venture capital commitments and the largest number of investments. At the same time, investments in biotechnology, electronic data processing (hardware and software) and environmental technology play a considerably smaller role. In fact, not all of the venture capital made available to US industry goes into high-technology sectors. The Federal Express mail service company is good example of a large investment which is not concerned with high-tech. Compared to figures for the US, Table 15 exibits the more traditional investment strategy in the case of German VC funds.

Table 15 German Venture Capital Industry Distribution
by DM invested*1994

Industry	Total Invested in DM million	%	Number of Companies	%
Agriculture	1,90	0,13	2	0,29
Chemistry	67,45	4,66	31	4,46
Stone/Earth	4,00	0,28	3	0,43
Iron/Steel	79,93	5,52	51	7,43
Mechanical Engineering	364,64	25,17	76	10,96
Electrical Engineering	41,06	2,83	54	7,83
Electronic Data Processing	44,70	3,09	56	8,09
Biotechnology	49,14	3,39	25	3,64
Environmental Technology	12,15	0,84	12	1,73
Precision Mechanics	3,40	0,23	12	1,76
Wood/Paper	76,39	5,27	36	5,18
Leather/Textiles	54,70	3,78	19	2,75
Food	79,88	5,51	22	3,13
Construction	61,14	4,22	48	6,94
Trade	204,73	14,13	98	14,16
Traffic	8,65	0,60	3	0,43
Communication Engineering	35,04	2,42	13	1,88
Financial Services	10,13	0,70	7	1,01
Others	149,86	10,35	124	17,91
Sum	1.348,92	93,12	692	100,00
not labeled	99,54	6,87	-	-
Total Sum	**1.448,46**	**100,00**	**692**	**100,00**

* Gross Investments, i.e. first-round and follow-up investments

Source: *BVK*, 1994

3.3 Framework Conditions and Development Determinants

United States

Taxation and Legal Aspects

As mentioned above, the growth of venture capital in the early 1980s coincided with the timing of the capital gains tax reductions of 1978 and 1981. New capital commitments increased from an average of $380 million for the period from 1976 to 1978 to an average of $1.01 billion for the period of 1979 to 1981 and to $3.93 billion for 1982 to 1984.[63] However, since the 1986 Tax Reform Act which raised the tax rate from 20 to 28 %, venture funding has declined, and then recently increased sharply.

However, Poterba notes that neither of these observations provides conclusive evidence for an inverse correlation between capital gains tax rates and venture capital funding.[64] It is true that the capital gains tax rate alters the relative return on investments for individuals. Yet one should note that taxes are only due upon realization of gains, which may not occur until many years after these gains accrue. The government in effect provides investors with interest-free loans on unrealized gains. But individuals account for a minority of venture capital funding. Institutional investors, such as pension funds, are typically tax-exempt. Investments through corporate subsidiaries face corporate tax rates. While changes in the capital gains tax rate may affect the cost of capital of the parent firm, these venture capital investments are much less sensitive to changes in the capital gains tax rate than investments through independent venture partnerships.

---

[63] Cf. Poterba 1989, p. 49

[64] Loc. cit.

Fund Construction

One of the many responsibilities of the general partner is regularly to raise new funds. In order to invest in portfolio companies on a continuous basis, managers must raise new partnerships before the funds from the previous partnership are fully invested (usually once every three to five years). Fund raising involves presentation to institutional investors and their advisors. Managers will first turn to the limited partners that invested in the previous funds in order to minimize expense.

A proven track record can make it easier for a venture capital firm to attract investors (see Table 16). The most commonly cited measure of a fund's performance is the internal rate of return (IRR).[65] However, returns are highly variable, and an impressive performance may reflect only one exceptional investment and not the overall picture. Therefore, investors may conduct an analysis of all but the highest-yielding investment of the partnership. Investors also look at the qualitative aspects of the venture capital firm. This includes investigating which of the general partners were responsible for managing the most successful investments. The gathering and analyzing of information is time-consuming and costly. Many institutional investors have come to rely on outside investment advisers to perform this service for them.

Table 16 Capital Commitments by Experience Level 1994 ($4.005 billion raised) in the US			
Years	**< 3**	**3 to 4**	**5 to 9**	**10 >**
Total Funds (%)	4,8	1,4	17,8	76

Source: Venture Capital Journal, February 1995

---

[65] See for a definition p. 8

The agreement includes certain provisions designed to protect the interests of the limited partners. The compensation structure, in which the venture capitalist receives a percentage of the fund's profits, provides an incentive for the fund manager to generate profits. There are also covenants, written into the partnership contract, which place restrictions on the activities of the general partners. The general purpose is to restrict the amount of risk that the general partners can undertake.

These covenants may establish a limit on the percentage of the fund's capital that can be invested in a single venture or on the total size of the two or three largest investments. They may also prohibit investments in publicly-traded and foreign securities, derivatives, or even other private equity investments that do not meet the partnership's specific focus (such as LBO investments). Covenants also may limit the amount of debt the fund can use, and may require that realized profits be distributed to limited partners immediately.

General partners also may make certain investments from which they would profit through fee income, or investments in companies in which their previous funds have equity positions in order to increase the valuation of these companies. Many covenants address these issues by limiting deal fees in restricting co-investments with other funds.

Another means by which limited partners can protect their interests is an advisory board. These committees usually consist of only the largest investors and they have very minimal powers of oversight. The board votes on matters of deal fees, conflicts of interest transactions, and other issues covered by the covenants. Some partnership agreements may grant limited partners the ability to terminate a partnership or remove a general partner from the fund.

Managing the fund involves several stages. The general partners first select investments. While investing in start-up enterprises is inherently speculative, venture capitalists take certain risk-avoidance measures. Perhaps most important is diversification across regions and in different financing stages. Co-investing, or syndicating, with other venture capital firms is also a method of reducing risk. A venture capital firm often obtains information about investment opportunities through investment bankers, brokers, consultants, lawyers, accountants, and managers from previously funded ventures. Professional agents may also provide deals for a fee. All proposals are screened, and those that meet the basic requirements of the fund (type of investment, industry, location, etc.) proceed to a more in-depth investigation to verify and scrutinize the details. Only about 10 to 30 % of the proposals pass the first screening.[66] Steps are then followed by a much longer due diligence process. The venture capitalist is interested in becoming more knowledgeable about the industry and the specific firm. For established firms, this means meetings with the firm's management, employees, customers, suppliers and perhaps consulting lawyers, accountants, and industry consultants.

The second stage is structuring the investment. This involves the negotiation of an investment agreement between the partnership and the firm. The structure details the financial and governance aspects of the deal.

The main financial issue is the amount of ownership that the partnership is entitled to in exchange for its investment. The main governance issues are managerial incentives for the portfolio firm, and the amount of control that the partnership will have over the firm.

General partners often sit on the Board of Directors, which is ultimately responsible for the management of the company. Among venture-backed biotechnology firms, general partners hold over one-third of the board seats.[67] Even as minority investors, venture capitalists will usually occupy at least one board seat in a

---

[66] Cf. Fenn et al. 1995, p. 3-26

[67] Cf. Fenn et al. 1995, p. 3-26

portfolio firm. The board is in charge of hiring and firing of the CEO and the monitoring of the firm's performance.[68] In addition, the general partners offer assistance by arranging additional financing, recruiting board members, and even consulting the firm on operations and strategies.

On average, a venture capitalist will visit a portfolio firm for which it is a major investor 19 times a year and spend more than 100 hours with each firm.[69] General partners often remain active in the portfolio firm until the distribution of stock to the limited partners. The partnership, depending upon the size of its voting block and the control rights associated with the shares it holds, has an influence on choosing from among the company's possible exit strategies: a public offering, a private sale, or a share repurchase by the company. A public offering generally generates the most returns for the partnership. However, a private sale has the advantage of payment in cash or marketable securities, whereas in the event of an IPO, existing shareholders are usually restricted from selling their shares for a certain period of time. The management of the portfolio firm may not welcome an acquisition since it often means that it loses its independence to the larger firm. A share repurchase by management is considered a last resort, and is mostly used when the investment has been unsuccessful.

The most common form of the investment is convertible preferred stock. In Germany, however, most common are direct investments and silent partnerships. This entitles the venture capitalist to prior claims both to the company's earnings, and, in the event that the venture fails, to the distribution of liquidated assets. The entrepreneur typically only can receive payoffs from the firm - e.g., dividends - after the venture capital fund has exited the investment.

---

[68]   Loc. cit.

[69]   Cf. Barry 1994, p. 5

This works in favor of the venture capital fund's investors for several reasons. First of all, preferred stock transfers risk from the venture capitalist to the entrepreneur by giving the former a superior claim on cash flows. The entrepreneur is also forced to prove his commitment to the project by agreeing to receive a percentage of the profits as opposed to a flat salary. This also provides an incentive for the entrepreneur to continue to work productively on the project. Staged capital works in a similar way to keep the entrepreneur working to see the company succeed.

The Role of Other Capital Suppliers or Financing Sources

Venture capital makes up only a fraction of the private equity market which includes all securities that are exempt from registration with the Securities and Echange Commission (SEC). Several studies in recent years have found that the majority of fast-growing companies started in the past decade have relied heavily on self-financing (through personal savings, credit cards, second mortgages) and less on venture capital funds.[70] One may distinguish between the organized and unorganized private equity markets. In the former, investments in the unregistered securities of private and public companies are professionally managed, as in the case of venture capital. Within the latter market, "business angels" are a critical source of seed capital. Angel capital includes investments made by wealthy individuals in small, closely held companies. Unlike venture capitalists, these individuals rarely take an active management role in the company. While they may act as advisors, they do not exercise direct control in the company. These partnerships are often arranged by lawyers, accountants, and other informal means. Investing directly may be the only option for investors in nations with only informal capital markets.

---

[70] A study by Amar Bhide (1992) surveyed 100 high-growth companies started between 1982 and 1989, and found that 75% were primarily self-financed. A Price Waterhouse (1992) survey of 700 nationwide software companies found that 80% have been self-financed.

The role of private individuals as a source of equity for new firms is illustrated by a study of 284 new technology-based firms founded in New England between 1975 and 1986. The study found that "private individuals and venture capital funds play complementary rather than competing roles in the financing of NTBFs".[71] While venture capital funds invested almost 5 times the amount invested by private individuals, 70% of the firms raised one or more rounds from private individuals, compared to 51% from venture capital funds. Individuals were the primary source of funds for deals under $1 million. Private individuals and venture capitalists provided roughly the same amount of seed financing. The role of private individuals declined dramatically with each successive stage, while venture capital funds were the predominant source of financing for the following stages. Though this study focuses on a small regional sample, it provides some insight into the interaction of venture capital and private investors in the equity market. The study concludes from their dominance in the seed and start-up stages, that private individuals "appear to have longer exit horizons and less risk aversion" than venture capital firms.[72]

Since the 1960s, the participation of large corporations in the venture capital market has followed the general cycles of independent partnerships. Corporations entered the venture capital market in the late 1960s, attracted by the well-publicized success of venture capital funds. Many large companies organized their own venture programs, investing either in venture capital funds or directly in start-up firms. Certain corporations began "internal" programs in which they attempted to foster innovation with the support of the corporate resources, such as financial, legal, and marketing support. When the IPO market declined dramatically in 1973, corporations followed the general trend in the venture capital market by scaling back on their own investments and programs.

---

[71] Cf. Freear /Wetzel 1992, p.77

[72] Ibid, p.89

The early 1980s saw a rebound for the venture market, and corporations renewed their efforts to take advantage of the potential rewards. In 1986, corporations accounted for around 11% of the total capital under management (see Table 1). After the 1987 stock market crash, and the subsequent decline in the IPO market, corporate venture programs were scaled down again.

While the number of corporate venture programs has generally followed the trends of the independent sector, there are three inherent problems which account for the lack of participation of corporations in venture capital. First, these programs have generally been established with multiple and, sometimes, conflicting objectives. A corporation may want the fund to focus on emerging technologies while at the same time generating attractive financial returns. These goals are rarely consistent. A second problem is that these programs have suffered from insufficient support within the corporations, as they are often considered by some managers as unnecessary luxuries, and often become expendable in times of financial hardship. And lastly, corporate venture funds have not succeeded because of their inability to attract talented managers. Unlike independent partnerships, corporate programs typically do not provide the 20% split of the profits to the fund managers, and often harshly punish failure. In addition, while independent venture funds acknowledge unsuccessful investments as an unavoidable part of the business, capitalists often continue to invest in failing ventures for fear of reputational repercussions of a failure.

Germany

Although the German capital market has shown a high growth rate over the past few years, the field of early-phase-financing of NTBFs is still underdeveloped. The following contains a discussion of the development problems on the German venture capital market. In this discussion, the tax system, the legal investment conditions for investors, the stock exchange, the alternative financing possibilities for firms as well as socio-economic factors are analyzed in the light of similar areas in the US.

Spread

Spread constitutes the main point of orientation for all financing and investment decisions owing to its impact on the choice of financial funds for firms and the attractiveness of investment possibilities.[73] Three factors have to be considered in relation to the tax evaluation of share financing: investors, investment shares and portfolio firms. The monetary flow among these different factors is shown in Figure 1.

The tax treatment of deposits by investors in investment shares depends upon the form of the deposit. If it is a share from a capital investment[74], it is subject to cooperation tax. If the deposit of the investor assumes the form of a share in a GmbH, OHG, KG, or if it is an atypical silent participation, it is subject to joint-venture tax, i.e. it is not subject to investment tax.

Figure 1   Monetary Flow of Share Financing

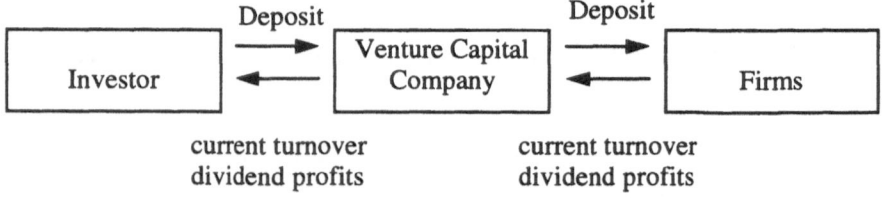

current turnover
dividend profits

current turnover
dividend profits

---

[73]   Analyses from the US, Great Britain and Canada prove that a negative correlation exists between the amount of "capital-gain-tax" and the investment in venture capital cooperations; see e.g. Poterba 1989; McMurty 1986; Eisenhardt/Forbes 1984.

[74]   For example shares: Assets of a joint-stock company GmbH cooperation rights of silent shares in a capital cooperation.

German law does not permit preferential tax treatment of investors in investment shares. The capital gained is subject to wealth tax.

The deposits of the investment shares in portfolio firms are subject to the same law as that applied to investors' deposits. Viewed from the point of view of investment shares, these deposits represent a pure re-allocation of wealth and are thus not subject to preferential treatment. The sale of investment shares and the conclusion of silent partnerships are of no consequence for portfolio firms. However, a company does face disadvantages in this case since, in contrast to the acquisition of long-term debt, the bases of assessment for trade tax now also includes 50% of the permanent debts on trade capital and 50% of the interest on trade earnings. As a result, private capital and profits are subject to a higher tax rate than foreign capital and the interest on foreign capital.

The taxes on the current returns from shares in a investment participation also depend on the type of investment. Those involved in capital investments are compelled to give investors positive or negative information (on capital assets) which is relevant for the income tax category. In this case the cooperation tax and the witheld capital gains tax are allowed as credit. This applies to holding companies since, as incoperated firms, they are considered as independent taxable entities. In the case of joint venture shares in a investment participation, the shareholders pay the tax. The shareholders can either take the tax receipts of the firm as profits or make losses valid.

The dispositions of investors' shares in an investment participation are taxed as follows: The disposition profits and losses of a natural person resulting from a share in a capital investment are irrelevant with respect to taxation. A share exceeding 25 % comes under a subsidized tax category as per Section 34 of the EstG. As far as joint venture investments are concerned, disposition[75] profits are subject to income tax. In this case, natural persons, for example single persons, only have to pay half the average tax contribution. Legal persons, for example stock corporations, are fully taxed regardless of the form and volume of share disposition profits.

The current returns on the investment shares from investments in portfolio firms as well as the disposition of these shares are subject to the same taxation regulations as those for current returns and disposition profits as far as losses from shares of legal persons in a investment share are concerned. From the perspective of a portfolio firm, current dispositions (as regular turnover usage) as well as the disposition of shares from investment shares are not subject to taxation.

To summarize, it can be stated that the German tax law does not provide for special regulations permitting the preferential treatment of investment shares which are financed through funds.[76] Tax advantages can however be derived from making use of the possible applications of tax laws; i.e. by the choice of a law form adapted to the typical ideal share-financing process. The establishment of a legal company (e.g. GmbH & Co. KG) in the early phases of investment share development, in which losses result from shares, provides investors with an opportunity to obtain tax waivers. Should the investment share fall in the "profit zone", the transformation into a stock corporation holds more advantages, since it is then possible to avoid the disadvantages of disposition profit taxation by disinvesting.[77]

---

[75] A tax reduction with respect to disposition profits by a joint venture is possible under § 6b EstG (Reinvestment possibilities)

[76] Cf. Wupperfeld 1993

[77] For the taxation application possibilities of share financing see Fischer 1987.

The question as to how far German taxation law, in comparison to taxation law in the US, hinders the development of the venture capital market cannot be answered in a satisfactory manner at this point. Nevertheless, there are some analyses available on this subject, which show that the German venture capital market does contain tax disadvantages when compared to the US.[78] These analyses, however, have a fundamental shortcoming in that:

- they only take certain aspects of the present the US and German taxation systems into consideration (e.g. depreciation possibilities, which reduce profits);
- they do not take into consideration the taxation of alternative investment possibilities for investors;
- they are based on the very convenient taxation law form combination in the US SBIC/Subchapter-S-Corporation, which - in practice - plays a negligible role.

The Statutory Investment Conditions for Institutional Investors

The statutory investment conditions for institutional investors regulate the extent to which specific groups of investors are allowed to invest funds, entrusted to them by third parties, in participations and VC companies. The statutory conditions hence also have an influence on the capital flow to such companies. What follows is thus an analysis of the opportunities open to pension funds, insurances, open investment funds and banks to invest in venture capital funds.

In the literature on this topic there is clear agreement about the reason for the sporadic upturn in the US venture capital industry in the eighties, particularly in the case of pension funds.[79] In the Federal Republic of Germany, however, pension funds do not play such an important role. This is because the liquidity reserves for

---

[78]   See for instance Nevermann/Falk 1986; Swoboda/Zecher 1985; Weichert 1987.

[79]   Cf. Bygrave/Timmons 1992

the provision of old-age benefits generally remain in the same firm as pension stockpiles for the purpose of preferential tax treatment

A further potential investment group, the insurance companies, have been able to invest 5 % of the stock wealth coverage and 6,25 % of the remaining collective wealth in the form of share capital since 1987.[80] So far, however, this increase in investment opportunities has only lead to a small degree of intensive involvement in investment-share-financing by firms in the insurance sector.

The structuring of investment shares as open investment funds could hold the advantage of greater functionality. This is not possible in the Federal Republic of Germany, however, since only those fund-supplying firms with access to the stock-exchange are allowed to pursue such investment policies. With the law aiming at improving the framework conditions for institutional investors that took effect as of January 1st, 1987, the legislative tried to make it easier for capital investment companies and insurance companies to acquire investment shares. However, this has had not yet had any noteworthy impact on venture capital market.[81]

As a consequence of the above, the banks in the Federal Republic of Germany constitute the most important investment group. This is the most important difference to the US, where it is illegal for commercial banks to pursue banking activities outside the banking sector and to invest in shares.[82] One limitation on the involvement of the German banks stems from their insignificant private capital quota, which does not allow for huge losses, e.g. from investments in shares with

---

[80] This includes shares that do not have access to official trade or are not considered in regulated auction, GmbH- or KG-shares, shares as silent partners as well as bonds.

[81] Cf. Fetzer 1990

[82] In the USA business banks are subject to the restrictive legislation of the Glass-Steagall-Acts from 1932, which was a reaction to the stock-exchange crash of 1929 and the resulting bank collapses that followed. Recently there has been an evident softening-up of the separation of banking systems. As a result since 1988 investment banking and since 1989 the stock market, has been partly legalized for banks.

high risks. Furthermore, share capital - together with other long-term investments - can only be granted up to an amount equal to that of the private capital. This is the reason behind the careful and risk-averse bank policies, which are referred to as prudent banking.[83]

On the whole, and independently of these limitations, the opportunities for institutional investors to invest in venture capital funds in the Federal Republic of Germany are by no means exhausted. Thus, state regulations do not have a very negative effect on the development of German share capital or the venture capital market. This means that institutional investors can, to a larg extent, continue to invest in VC participations as they have been doing up to now. However, a greater relaxation of investment regulations as a whole could lead to an innovative and diversified financial sector in the long-term, as well as serving as a basis for the rapid acceptance of venture capital in the financial sector.[84]

The Stock Exchange

The structure of both the stock exchange and, in particular, of access regulations for specific segments of the stock market have a direct influence on how, and at what prices, and stocks can be quoted. Consequently, they also have a direct influence on the dividends which can be harvested by VC. The empirical analysis of the US confirms a strong correlation between the development of the OTC market[85] and the targeted dividends of investment shares, and demonstrates that without the exit

---

[83] The German banks are however in international comparison relatively risk orientated; see Büschgen 1985; Quillmann 1987; Stedler 1993.

[84] Cf. Wupperfeld 1993

[85] The American stock exchange market consists of AMEX (American Stock Exchange), NYSE (New York Stock Exchange), NASDAQ (NASD Automated Quotations System) and the OTC market (Over The Counter), which is regarded as the most important exit option for share cooperations; see for an overview chapter 2.3.

option of the stock exchange, venture capital in the US would not have its current economic capacity.[86]

In the Federal Republic of Germany, the trading of stocks and shares takes place officially on regulated markets and on the free market outside the stock exchange. The free market is not subject to the rules of the stock exchange, but operates solely on a private basis regulated by the law. However, it only enjoys a very limited degree of acceptance, due the open question concerning liability, among other things. The official trade is particularly geared to large and established firms as well as to federal bonds and other emissions.

Due to its highly restricted access, the regulated market which came into effect in 1987 represents a hurdle with respect to the age of firms, the minimum emission volume, as well as spread and access fees[87], which are convenient for the disinvestment by investment shares.[88] This exit option has, however, not played a significant role for German investment shares up to now. Fetzer makes the universal banking system in Germany and the market power of the banks responsible for this development. The universal banking system allows credit institutes to function as emission partners, a fact that represents a difference to the new US emissions, which almost always rely on the participation of the banks.[89] Since the provisions of small firms are relatively small, and they operate in a narrow market that makes it difficult to control trends, the attractiveness of emissions declines as the size of firms becomes smaller. Furthermore, being liable for the prospects of young and small firms is a high risk for the emission partner. Credit institutes therefore prefer

---

[86] Cf. Bygrave/Timmons 1992

[87] The emission volume must amount to a minimum of DM 500,000 (official trade DM 2,5 million), the required minimum of permanent stock in three years for official trade and one year for the regulated market.

[88] Cf. Stedler 1987

[89] In 1983 the USA had 300 emission institutions, the Federal Republic had three, see Weichert 1987.

to finance small firms with loans, which is also largely due to the absence of competition in the field of private capital mediation on the stock exchange. It is only after a specific critical level has been reached that the banks turn their attention to the stock exchange. This business strategy is kept secret due to the strong minimum requirements to which potential emission firms have to adhere.[90]

Due to the fact that the organized capital market in the Federal Republic of Germany has not operated as a "disinvestment option" for shares in NTBFs, there is no pull effect which could have also helped the development of an accelerated capital segment on the venture capital market. On the whole, it is quite apparent that the absence of exit options on the stock exchange has narrowed down the dividend opportunities open to investment shares in NTBFs. Hence, it provides framework conditions that have hindered the development of the seed capital market. As a result, German investors in VC plan to regulate stock exchange access for NTBFs on the US-American computer stock exchange, the NASDAQ.[91]

Alternative Financing Possibilities for NTBFs

It is obvious that the NTBFs' demand for share capital is dependent on the financial alternatives available to them. What follows is thus a discussion about the 'rival products' , e.g. long-term bank loans, investment promotion funds as well as private venture capital and their effects on the financing behavior of NTBFs.

Medium-sized firms searching for capital in the Federal Republic of Germany have several highly liquid financial sources in the form of foreign capital, which is largely made up of long-term bank loans. The banking sector in the Federal

---

[90] Cf. Ingram/Miles 1984; Hausberger 1984

[91] The TVM Techno Venture Management GmbH & Co. KG plan for 1995, the stock exchange access for Spea Software AG in Starnberg, der Dalim GmbH in Willstätt, and the Qiagen GmbH in Hilden; see Ludsteck 1994.

Republic of Germany plays a dominant role in firm financing,[92] so that German industrial investment companies regard the bank sector as a relatively strong competitor, in contrast to the situation in the US. German credit institutes are more willing to take risks than their US counterparts. For example, their requirements determining the volume of private capital in the case of granting credit are considerably lower in the Federal Republic of Germany than in the US. This is a result of the higher savings by private households and the good coverage of financial markets which, together, favor high bank liquidity and foster a greater willingness to grant loans.[93] In addition to this, the banks have a new orientation, intended to increase their performance horizons, through consultancy services,[94] for example, and departments for financial innovations.

As a result, the demand for share capital is lower in the US. Fetzer even associates this with a surplus of supplies of foreign capital, which has reduced the demand for private capital.[95] His analysis is not, however, explicitly limited to the financing of NTBFs in their early development phases, but focuses more generally on medium-sized firms. The latter are confronted fewer hurdles in the acquisition of foreign capital, whereas banks tend to be reserved about granting credit to very young high-technological firms where the risk association is greater. His report however remains valid, due to the dominant role of the banking sector in financing firms. On the whole, this hinders the development of venture capital markets, and has, in turn, led to the development of a share capital culture.

In contrast to the US, there was an express willingness to support young and medium-sized firms in the old Federal States of Germany until a few years ago. Under the framework of the support programs for small and medium-sized

---

[92] Cf. Lucas 1994

[93] Cf. Schramm 1988

[94] For the state of development of bank consultancy firms, see Rüschen 1990.

[95] Cf. Fetzer 1990

companies organized by the Kreditanstalt für Wiederaufbau (KfW) and the Deutsche Ausgleichsbank (DtA), the invested funds amounted to more than DM 8 billion in 1988 alone.[96] In the eighties well over 300 NTBFs were supported by the pilot scheme "Technologieorientierte Unternehmensgründungen" (TOU).[97]

The effect of these support measures on the demand for share capital calls for differentiated judgment. The offers to support young and medium-sized firms have, on a whole, surely tended to hinder the development of the venture capital market. This does not, however, explain the minor importance of seed segments in relation to the venture capital market as a whole. Support can even have the opposite effect in the case of NTBFs, especially when it triggers the creation of innovative firms, thus making such firms interesting investment objects. This means that such support has a "finance-postponing" function, since technologically oriented firms are put in a position where they can carry out development work, develop a technological basis and finance their first marketing activities. Once they have reached this development phase, NTBFs are regarded as a reduced risk by VC investors, especially due to the fact that these firms now have larger capital basis.

An important financial source for the seed phase of newly created firms in the US is represented by informal share capital. These funds are made available by family members, friends or independent private persons (business angels). As opposed to the US, this type of "seed culture" is practically non-existent in Germany.[98] In addition to this, any such involvements in the Federal Republic of Germany are

---

[96]   Cf. Müller-Kästner 1989

[97]   The pilot scheme TOU was started in 1983; the deadline for applications was 31.12.1988. 319 JTUs were supported in the Support Phase II (subsidies for R&D up to DM 900,000). For goals, instruments and results of the scientific reports accompanying the pilot scheme TOU, see Kulicke u.a. 1993. Support under the framework of the TOU pilot scheme can with respect to the innovation process, be subdivided into three phases (conceptional, R&D, and market introduction), for which the total expenditures, of up to DM 800,000 for the development of an innovative product or process are to the most part support subsidies, see Bachelier 1993.

[98]   For a description of the German situation see Frommann 1993; about business angels in US and Great Britain see e.g. Harrison/ Mason 1991; Wetzel 1983; Wetzel 1987; Mason/Harrison 1992.

regarded more as capital investments and less as an intention to participate in the management of a firm. As a result, many NTBFs in Germany, whose founders neither have sufficient private funds nor the ability to acquire means of support, lack this form of postponement financing, which can also include supporting a firm by evaluating its concepts.

To summarize, one can say that it is not possible to precisely define the effect of alternative financing funds for the development of the venture capital market. On one hand, the predominance of the banks in financing these firms and the comprehensive support-programs for medium-sized firms surely hinder the development of the venture capital market as a whole. On the other hand, the support for NTBFs ought to have a stimulating effect on the whole.

3.4   Public Policy

United States

The venture capital process is sensitive to a wide variety of government policies, most notably taxes and security regulations, that alter the reward to riskiness of various investment opportunities confronting investors. A report to the US Congress' Joint Economic Committee conducted a survey of 277 leading venture capital firms in the US.[99] The Committee suggested several public policies to improve the entrepreneurial climate. These include funding for basic research at universities, research and development tax credits for commercial enterprises, and antitrust regulations to encourage formation of joint research and development ventures among firms.

State governments also play a role in venture capital. Venture capital activity is highly concentrated in three areas of the country: California, Massachusetts, and

---

[99]   Cf. Premus 1984

New York-New Jersey. Capital disbursed in 1994 in California was almost four times that of the next highest state, Massachusetts. This "regional gap" means that entrepreneurs in other areas are at a competitive disadvantage in obtaining this source of funding for their ventures.

This disparity of capital across industries, regions, and stages of development, as well as an increasing trend toward later-stage, less risky investments, lead some to believe that important technologies that could provide long-term economic and social rewards have been neglected by the equity markets. These claims have renewed the debate over what role, if any, the US government should play in the financing of entrepreneurs. While most economists and policy-makers agree that the government should encourage new ventures, there are conflicting theories as to whether the government should act directly as a venture capitalist. Anthony Clark, is a member of the House Committee on Science, Space, and Technology, who, like the current Clinton administration, supports a partnership between government and the venture capital industry. Clark suggests that public funds should be directly invested in those ventures that would promote innovative technologies.[100]

The economists, Florida and Sahlman,[101] argue that the government would be a poor venture capitalist. Public interests are often incompatible with the high-risk, high-return world of venture capital. Two previous programs show that government backed programs have not been successful: the Small Business Administration's SBIC program and the many state venture capital programs. The rise and fall of the SBIC program has already been discussed. During the 1980s, a number of states created venture capital funds, hoping to foster high technology enterprises, for example Florida, Michigan and Washington. Essentially, these states were attempting to recreate the Silicon Valley phenomenon in their own area. In 1990, 23 states had committed a total of $200 million in public capital for their funds. These

---

[100] Cf. Clark 1994, p.49

[101] Cf. Florida 1994; Sahlman 1992

programs have essentially failed. Many of the investments did not pay-off, or the capital went to other states with better investment opportunities than Florida. Similarly, Sahlman argues that there is no shortage of venture capital in the United States and that the United States possesses a strong financial system, which is continually creating new sources of capital.

Individual states are consequently focusing on increasing the number and quality of entrepreneurial companies in order to attract venture capital to the region. States differ among the quality of education they offer. Universities are one factor behind the regional gap. Cambridge, Massachusetts, for instance, is the home to both M.I.T. and Harvard University, and also has a large number of software companies. University research provides innovation and the skilled labor force that is necessary for the application of new technologies and the commercial development of new products and processes. Technology companies in particular require highly skilled employees.

In general, academics concur with the venture capitalist. While both groups support the idea of government programs such as grants for education and research, the majority oppose direct government involvement in the venture capital market.

Germany

State support-programs can also be added to the framework conditions relating to the development of the venture capital market. In the Federal Republic of Germany, the pilot scheme "Beteiligungskapital für junge Technologieunternehmen" (BJTU, share capital for young high-technology firms) has substantially improved the conditions for investing in NTBFs. This scheme was established in July 1989 because of the insufficient development of the seed capital market and was scheduled to end in December 1994. The pilot scheme was intended to encourage leading investors, finance the early development phases of NTBFs and contribute towards the development of the NTBF risk capital market, which, in the medium-term, largely exists without state support.

This pilot scheme offers two modes of access involving divergent support conditions. In the first mode of access, the Kreditanstalt für Wiederaufbau (KfW, the Reconstruction Loan Corporation) grants refinancing loans in exchange for a 40 % share in the profits made by the leading investors, who in turn invest this capital in NTBFs. In case of default, the leading investors are normally only liable to the extent of 10 % of the refinancing amount. In the other mode, the co-investment approach, the Technologiebeteiligungsgesellschaft (Tbg, Technology holding Company,), a subsidiary firm of the Deutsche Ausgleichsbank (DtA, German Compensation Bank), only participates as a silent partner in a NTBF, if a leading investor is also involved with the same sum. Within a period of three years after the investment participation has been terminated, the leading investor can offer his share to the Tbg at a lower price, or buy the Tbg's silent partnership, in which case he must also pay an additional charge.

The BJTU pilot scheme has resulted in some of the classical investment companies, in particular the MBG, stepping up their investments in NTBFs again. At the same time, new suppliers of risk-bearing capital appeared on the scene. Among those worth mentioning here are the newly founded seed capital investment companies and a number of credit institutes. A total of approximately 40 % of all investment companies on the German market took advantage of the BJTU pilot scheme and had invested around DM 300 million by the end of January 1995 in support of share capital; they were involved in 371 support commitments to NTBFs.[102]

---

[102] Cf. Kulicke 1995

3.5   Specific Conclusions of the Literature Review

The results of our literature screening led to a very heterogeneous picture with regard to the US and German VC markets. This summary shows that one can note predominantly positive development factors for the US:

- there is a relatively long tradition of VC, which began in 1946 with the foundation of ARD; this development has lead to large number of highly professional VC managers;
- policy schemes such as the implementation of SBICs in 1958 paved the way for broader involvement by venture capital institutions;
- acts and regulations such as the ERISA "Prudent Man" Rule, the Small Business Investment Act, and the ERISA "Safe Harbour" Regulation have created a stable framework for investors and investments and made available a growing supply of money for VC activities.

Developments such as technological breakthroughs in microelectronics and biotechnology and acquisitions of NTBFs by large Japanese as well as US companies are also considered to have had a beneficial effect on the growth process of venture capital in the United States.

If the development of the German venture capital market is taken into account, one can say that the situation for VC and NTBFs has obviously become more favorable thanks to public policy, especially the support provided by the BJTU pilot scheme and the measures it has taken, for instance, the BTU and TOU pilot schemes. The NTBF venture capital market would have been far less significant without such a massive support. When considering German framework conditions, we mainly discussed the development hurdles which allegedly hinder the development of the venture capital market here. Among these the most important are:

- the universal banking system in Germany and the dominant role of credit institutes in financing young and small firms;

- the lack of disinvestment possibilities available to firms that want to go public. IPOs in particular play a minor role in comparison to the opportunities offered by the NASDAQ or the pertinent entry and valuation regulations. Those most effected by this situation are the younger and high risk companies. Hence, capital investment companies lack one of the preconditions for realizing high dividends from their investment in NTBFs;

- the absence of preferential tax treatment for share capital.

In the Federal Republic of Germany, a long venture capital tradition does not exist as it does in the US. It was only in the eighties, for example, that there was any market volume worth mentioning However, most companies interested in investment participations focused on financing the growth processes of established firms, MBOs, etc. Due to the absence of a VC tradition and the relatively recent development of a venture capital market one can say that German VC industry still lacks experience compared to the US venture capital market. Up to now, only a few investment capital companies have had extensive and all-round experience with NTBFs. Efficient networks and suitable instruments for evaluating, managing and supervising the activities of NTBFs, which are necessary in order to harvest higher dividends, have consequently only been developed recently by a few specialist companies. In contrast to the US, a "business angel community" does not exist. However, it may be noted that public policy in Germany, as in the US, has provided some decisive impulses for the development of a VC market.

4   Ten Case Studies of Venture-backed Firms and their Investment Companies

4.1   Methodological Preliminaries

It is widespread knowledge among experts of research strategies that both qualitative and quantitative methods have their own value. At its best, a comprehensive research design would include both approaches in order to overcome the deficits of each strategy. However, there are reasons for deciding which strategy would be more appropriate given the proposed questions. Pragmatic reasons (the available research budget or time) as well as context-related reasons, often influence this decision, such as a very heterogeneous field of investigation, for example, in cases where the necessity of pilot or exploratory research is very great.

Much has been said about the value of exploratory studies as an antecedent to more extensive survey activity in the literature on research methodology.[103] Following the concept of investigation mentioned in chapter one, this study did not aim to carry out investigations using large postal surveys, but to focus on a more qualitative/exploratory type approach. This approach was adopted for a very heterogeneous field of investigation. The purpose of this exploratory study was to compare the US and German venture capital systems, focusing on new technology based firms as one of the most interesting vehicles for stimulating technological development. An empirical comparison of US and German venture capital ought to show, in particular, whether VC managers pursue diverse performance strategies

---

[103] Cf. Zikmund 1991, p. 75 ff.

and identify any country-specific differences from their involvement in NTBFs. Given the research sources available (of approximately six man months), ten case studies (five for each country) were undertaken in winter 1995/96.[104]

In arguing the relevance of a case study approach, Yin observes:

"In general, case studies are the preferred strategy when "how" or "why" questions are being posed, when the investor has little control over events, and when the focus is a contemporary phenomenon within some real life context".[105]

Yin goes on to suggest that such `explanatory` case studies can also be complemented with `exploratory` and `descriptive` approaches. The objective of the research, particularly its investigatory and descriptive nature, closely follows Yin's logic for case study methodology. Within the typology of case study approaches, the design used may be identified as a `multi-case, embedded` research method. The term `embedded` here refers to the duality of the units of analysis, namely the venture capitalist investor and the NTBF entrepreneur investee.[106] The design is distinguishes itself from the `holistic` approach which does not discriminate or focus on differing units but analyses the situation or the case per se.

As Mitchell has argued, the dominant influence of quantitative methods has meant that "...`representativeness` has come to mean typicality in the sense of a statistically reliable random sample from a population".[107] The purpose of the case study approach is, in contrast, to "expand and generalize theories (analytical generalizations) by a process of inference and "not to enumerate frequencies

---

[104] We discussed different approaches of case study research in the context of the empirical design of this study. Helpful in this context was the research design in the study provided by Murray 1995.

[105] Cf. Yin 1989, p.13

[106] See also Pettigrew's 1973 `triangulation` methodology, which uses multiple respondents.

[107] Cf. Mitchell 1983, p.201

(statistical generalizations)".[108] Thus, while the analysis of ten case studies arrived at a number of propositions, these cannot, on the basis of the employed research methodology, presently be attributed to the wider population of US and German venture backed NTBFs.

The sample selection was not random. The sample consists of companies - VC companies and NTBFs - that were familiar from previous research projects and called for further investigation. The criteria used to select the research respondents were based on a combination of pragmatic and context-specific considerations. One of these was concerned with the available budget and time.[109] In this context, another consideration was the synthesis between the Harvard Business School approach, which requires several months of detailed investigation in each case, and the German approach, which is based on shorter investigation periods. A third aspect was the clarification of existing rights of use and copyrights as well as providing companies with the opportunity to give interviews. It is obvious that ten case studies are not sufficient to cover all segments of the venture capital market financing new technology based firms in each country.

Chapter 3 discussed several types of investment companies within the German VC market. However, not all of them have a pendant on the US market. Instead of selecting pairs of US-German VC companies for comparison, the decision was made to present a broad and diverse picture for each country. Thus, although the case studies presented here are not representative, the empirical information they contain can be regarded as complementary to the findings of the literature review.

The US and German case studies will be used to illustrate different ways in which venture capital organizations provide management, oversight and control of

---

[108] Cf. Yin 1989, p.21

[109] The original concept was calculated on the basis of a larger grant from the German Marshall Fund. Owing to the cuts, it was necessary to make adjustments to the concept, especially the empirical investigation.

portfolio firms. The US cases illustrate the diversity of the venture capital industry: from the more traditional form of the start-up fund to the relatively new development of the venture leasing fund. Furthermore, the German case studies take into account the heterogeneity of types of investment companies as well as the special framework conditions for NTBFs in the new Federal Laender.

4.2   The US Studies

4.2.1   The Early Stage Investment : the Start-up Firm

Overview of Portfolio Company: AccessLine

AccessLine, a telecommunications firm based in Bellevue Washington, was established in early 1989. It had revenues of $14,495,883 for 1994, an increase from $8,359,789 the previous year. The company developed the "One Person, One Number" concept, which allowed individuals to manage all of their telecommunications (personal and business calls, faxes, voice messaging, and paging) through a single phone number. In order to reach a subscriber, a caller dialed an assigned number, and then AccessLine directed the call to the subscriber. Thus, AccessLine strategy was distinct from numerous manufacturers of personal communication systems who were contemporaneously attracting financing. While a variety of competitors were focusing on developing special units that were usable inside and outside the office, AccessLine sought to achieve similar goals using the customer's existing telephone network. The entrepreneur of AccessLine - a technological expert with business experience from his first company that was sold to a downstream manufacturer - believed that consumers would be much more willing to pay a few dollars a month to get most of theses communicators' features from AccessLine. In 1989, AccessLine entered into a strategic relationship with McCaw Cellular Communications. McCaw introduced AccessLine's product to the commercial marketplace and also provided a considerable fraction of the firm's initial financing. Over the next three years, the company licensed its service to other carriers and formed strategic alliances with equipment manufacturers. In July 1994, AccessLine undertook a $15.5 million private placement from five investors.

Overview of Venture Capital Company: Apex Investment Partners

Apex Investment Partners was founded in 1987. In its eight-year life, Apex had raised three funds. Its first fund, Apex Investment Fund I, had closed in 1987; Apex Investment Partners II, in 1990. In May 1995, Apex was in the process of raising its third fund with an intended commitment of capital of $75 million, of which a pension trust fund had already committed $10 million. Apex's last two funds had a combined capital pool of $70 million concentrated in four broad industries: telecommunications, information technology, and computer software; environmental and industrial productivity-related technologies; health care and related technologies; consumer products and speciality retail. As of mid-1995, the second fund, Apex Investment Partners II, had yielded a cash return of 26% to its limited partners.

General Aspects of Venture Capital Management Strategy

In its past two funds, Apex had sought to balance early-stage investments with investments in firms that were already generating revenues. At the end of 1994, 35% of Apex's portfolio was in start-up or first stage investments. 1/6 was in buy-outs or industry consolidations. In most investments, Apex sought the role of lead investor. At the end of 1994, Apex was the lead investor in 2/3 of its portfolio and in 3/4 of the investments, an Apex employee had joined the board.

In evaluating potential portfolio firms, Apex considered how that investment fit into its larger portfolio. Like most venture capital companies, Apex generally relied upon a few investments to offset its losses, and therefore it was important for the fund to be diversified in terms of regional location, financing stages, and industries. Apex followed the general guidelines for earning high rates of return - investing early, avoiding dilution, and picking companies with enormous upside potential. Ideally, the portfolio firm would have an exclusive right to a specific market, as through a patent or license. Apex also evaluated the people behind the entrepreneurial and management teams, as they would be an important factor in determining whether the venture will ultimately succeed. Another consideration in making investment decisions was Apex's syndication partners. The second fund's partnership agreement prohibited Apex from investing more than a few million dollars in any one deal.

The Relationship between the Venture Capitalist and the Portfolio Firm

AccessLine turned to the venture capital industry in early 1995, intending to raise about $16 million. In the proposed deal, Apex Investment Partners would contribute a total of $2 million of its own capital and arrange for additional financing (Apex and its affiliated investors would contribute between $4.5 and $6.5 million). Despite its six years of operation, AccessLine was considered an early-stage investment from the view of venture capitalists.

Apex conducted a relatively extensive evaluating process of AccessLine. This meeting and an earlier session had been spent establishing an understanding and a relationship of trust between the corporate managers and the venture capitalists. Three venture capitalists from Apex Partners and several analysts from an affiliated firm spent nearly 12 hours with representatives from AccessLine. AccessLine fit into Apex's objectives for its third fund in several aspects. AccessLine was a fast-growing, telecommunications firm, which had just invented a differentiating personal communication service. The company also required relatively modest capital and operating expenditures, and had a promising future. AccessLine had signed agreements with several important players in the telecommunications industry which allowed it to generate increasing fees as the use of these services increased. Through its relationships with European and Canadian carriers, the company had potential for rapid international growth . In addition, Apex considered AccessLine's management team to be an asset: it combined extensive technical expertise with business experience in many areas of the telecommunications industry, and the company had met or exceeded financial projections during the last seven fiscal quarters. Unlike most early-stage deals in which Apex considered investing, AccessLine had relatively limited risk : the company was likely to return a positive cash flow, possibly through an IPO, in the near future. This would balance Apex's concentration of investments in early-stage portfolio firms, which were only in the initial rounds of financing and far from being harvested.

Despite AccessLine's considerable merits, Apex remained hesitant on account of two issues: valuation and managerial incentives. The five private investors who had provided AccessLine with capital in 1994 had to be taken into consideration. AccessLine's management wanted Apex's valuation (Series B) to be priced at a premium to the valuation of the private investors (Series A).[110] AccessLine believed that the company's value had appreciated since the initial financing. Therefore, it would be unjust to the Series A shareholders to offer Apex a better deal. Series A preferred shareholders would have to approve the financing, and might ask to re-negotiate with AccessLine. Apex, on the other hand, was wary of investing at too high a valuation because it would limit the investment's potential returns. Apex also needed to receive the approval of its syndication partners.[111]

The second issue was that Apex Partners wanted the contract to include more direct incentives for AccessLine's management team to consider the interests of the

---

[110] In venture capital, successive financing rounds in which shares of convertible preferred stock are issued, are often denoted by 'Series A, Series B, etc..' Therefore, in the case of AccessLine, 'Series A' refers to the initial $15.5 million financing from private investors, and 'Series B' refers to the proposed financing agreement with Apex.

[111] The actual valuation was also complicated by the fact that AccessLine was not publicly traded. Therefore, it was difficult to value the deal using the Black-Scholes model. Black-Scholes is the standard formula used to value financial options, and can also be used to value warrants, although the formula does not reflect the illiquidity of a warrant. Apex proposed to be the lead investor in a $16 million Series B financing round. Each $8 share would consist of a preferred share and .7 of a warrant for an additional share. Apex wanted to have a seat on board. If the firm did not succeed in going public, the warrants would not have much value. The question remained as to how much of the unit price should be assigned to stock versus the warrant.

partnership. As lead investor, Apex Partners would occupy a board seat and, therefore, would have some influence on decisions about the means of exit. However, this did not guarantee that AccessLine would go public, or that the IPO would be made at an attractive valuation. Apex proposed several contract provisions that would serve as incentives in order to accomplish this objective: punitive interest of dividend payments if the firm did not go public; the right for the venture capitalist to fire management if the firm did not go public by a certain date; the ability for the venture capitalist to require AccessLine to repurchase their shares (as with a put option).

The proposed deal with Apex Partners would provide the capital AccessLine needed for future acquisitions and the development of strategic partnerships. Apex Partners intended to serve as lead investor for this transaction with AccessLine, and therefore would be granted a seat on the firm's board. Because this transaction was at such an early stage, it was too early to tell what kind of management relationship Apex would establish with this particular portfolio firm.

4.2.2   The Later Stage Investment: the IPO

<u>Overview of Portfolio Company: ImmuLogic</u>

ImmuLogic Pharmaceutical Corporation had revenues of $3.265 million and a staff of 72 in 1990. The company was formed in 1987 by Dr. Malcolm Gefter, a professor at the Massachusetts Institute of Technology (MIT) since 1977. In the early 1980s Gefter hired a group of researchers to commercialize products from his laboratory. Gefter had discovered a less expensive, less time-consuming, and safer alternative to allergy shots. As of February 1991, the company had 11 patents pending, and expected its first product, for cat allergies, to be ready for sale in 1995. Products for allergies to ragweed, dust, and grass were expected to be introduced after that. In 1986 and 1987, Dr. Gefter and MIT applied for patents, and in the meantime discussed licensing the technology with several venture capital firms. In

June 1987, three venture firms - Rothschild Ventures, Technology Venture Investors, and Greylock Ventures - provided the first round of funding for ImmuLogic, a total of $3.4 million. One principal from each firm joined the board. In 1988, the initial investors and four other venture funds undertook a second round of financing of $12.8 million. The price rose to $6 per share, from the initial financing figure of $2 per share, due to ImmuLogic's success during the past year.

In 1989, ImmuLogic joined with Merck, a major pharmaceutical company. This had several advantages for ImmuLogic as it would enhance the credibility of the start-up firm, bring ImmuLogic access to a major research facility, and provide ImmuLogic with additional working capital. The third financing, which included Merck and investors from the previous round, provided the company with another $12 million. The valuation of ImmuLogic increased from the second round valuation of $35 million to $65 million. However, ImmuLogic needed additional operating revenue as its products remained in their development stages. Venture capitalists, therefore, considered ImmuLogic to be a high-risk, early-stage investment.

Overview of Venture Capital Company: Greylock

Since its establishment in 1965, Greylock had established several partnerships with a total equity investment in excess of $300 million. These investments encompassed a broad range of industries, including biotechnology. Greylock was unique in the venture capital industry in that it maintained relationships with a small number of limited partners, consisting of individuals, families, and university endowments. These partnerships were also unusual in that they were long-standing. Another important distinction from other venture capital partnerships was the absence of pension funds as investors. These institutional investors, obligated to produce high rates of return within a relatively short amount of time for its clients, were often unwilling to enter into long-term investments. Therefore, due to the enduring and familiar nature of Greylock's relationship with its limited partners, the organization was able to invest in ventures which would sometimes require as long as ten years to show a profit.

General Aspects of Venture Capital Management Strategy

Henry McCance, a general partner at Greylock since 1969, believed that in evaluating portfolio firms, it was essential to make reasonably priced investments in firms with market-leadership potential, and with strong management teams. McCance also held that the venture capitalist should work as an active partner with management. In assessing a potential deal, McCance had to consider his own reputation, as one of greatest assets was his access to "deal flow". Because a venture capitalist relies upon his track record to attract future deals, McCance viewed ImmuLogic as one in a long string of transactions.

In evaluating an investment, McCance would normally be reluctant to have a company in his portfolio go public that was so far from profitability. However, he understood that the biotechnology industry was unique in that the firms that undertook initial and secondary offerings remained high-risk development-stage projects. Venture capitalists typically took biotechnology companies public in the early stages of the regulatory approval process, a long time before their products generated revenues. It might take over ten years to receive government approval.

This case highlights the importance of networks. A general partner from a California-based firm, introduced McCance to Gefter, believing that Greylock would be a valuable syndicate partner in the deal. In addition to its extensive biotechnology experience, Greylock was located in Boston and had close ties to MIT, a limited partner of Greylock. Therefore, because of its physical location, investment experience, and financial ties, Greylock could provide effective oversight in the deal.

Relationship between the Venture Capitalist and the Portfolio Firm

McCance considered the investment worthwhile in that ImmuLogic, while still a high-risk venture, was of a high quality relative to other new biotech firms. One of its greatest assets was Gefter. McCance met with Gefter and was impressed by his business acumen, communication skills, and scientific acuity. However, McCance was not confident that he could assess the scientific validity of ImmuLogic's concept. Therefore, he asked the CEOs of Greylock's previous biotechnology portfolio companies to assess Gefter's papers. McCance also traveled with Gefter to a conference in Colorado on T-cells. Here he assessed Gefter's standing with his scientific peers.

Biotechnology companies use several sources of public and private financing in order to raise the tremendous amount of capital required to create commercially viable products. Venture capital funds are the most significant source for early stage biotechnology companies, providing both financing and oversight. From ImmuLogic's perspective, a deal with a venture capital firm like Greylock could add value to its company through both financial industry contacts and credibility. McCance, in particular, had many desirable assets. Being responsible for a number of successfully established portfolio firms, McCance had established an important network of financial industry contracts. Second, he would bring credibility and experience to the deal. His prominent role in deals with several privately-held companies would induce investment bankers to treat ImmuLogic fairly.

Greylock's partners had developed extensive experience in the biotechnology industry by serving as a lead or a co-lead investor in two early stage biotechnology companies in which a representative had sat on the board. It had provided second-round financing for another biotechnology start-up. These transactions provided the partners with a valuable understanding of the specific management challenges posed by the industry. Perhaps more importantly, these three portfolio firms proved to be among the most successful biotech companies in the 1980's, earning Greylock a reputation as an astute investor in biotechnology firms.

McCance was unlike most venture capitalists in that his company generally held its shares until the fund expired. The Greylock fund which invested in ImmuLogic was set to run through 1995 and, therefore, McCance would remain a major shareholder and director at ImmuLogic until then. (In a few instances, a venture capitalist, such as McCance, might remain invested in and on the board of companies for a decade or longer.)

A lead venture capitalist such as McCance had a great deal of influence in a privately held firm as an IPO approached. In addition to the contractual terms that guarantee shareholders legal voting rights, the entrepreneur tends to defer to the experience of the venture capitalists. McCance, along with the general partners form two other venture capital companies, played a key role in taking ImmuLogic public.

In March 1991, McCance and Gefter believed that the time was right for ImmuLogic to go public. In commercializing its allergy product line, ImmuLogic was considering a partnership with an established pharmaceutical company. McCance believed that an IPO would greatly enhance the company's negotiating position by providing an infusion of assets. Another reason for taking ImmuLogic public at that time was the pressure of an IPO from ImmuLogic's closest competitor, Cytel Corporation. It was common for the first IPO from a specialized technology sector to attract a large amount of the institutional money which invested in that sector. Therefore, Cytel's public offering could put ImmuLogic at a disadvantage in its initial offering, or even its finding of private financing. McCance feared that ImmuLogic's IPO would be eclipsed by other biotechnology IPOs. In the past during periods of intense IPO activity, the best firms tended to go public early in the cycle. The IPO market was "hot" in 1991 and a number of firms still in preclinical trials indicated in preliminary filings with the SEC that they intended to go public.

There were also arguments against an IPO at the time. One of the investment bankers which the partnership consulted felt that an IPO was premature because ImmuLogic would be able to command a considerably higher valuation if it waited until it had begun human clinical trials. This might mean a valuation of $200 million instead of the $100 million estimated at the time. In addition, the danger of a withdrawn IPO would be considerably reduced. The failure to sell its shares would have a substantial negative impact on the ImmuLogic's reputation and might impede its ability to sell shares in the future. In spite of these concerns, ImmuLogic issued an IPO in 1991.

4.2.3   Corporate Venture Capital

Overview of Portfolio Company: Documentum

Documentum, Inc. reported $10.7 million in revenues for 1994, reaching profitability in the fourth quarter of that year. It projected revenues of more than $25 million for 1995. The company developed and marketed the Documentum Enterprise Document Management System, a family of open client/server software products that can be tailored to specific business-critical document applications. The Documentum System enabled the capture, work-flow, assembly, and distribution of business critical documents. Documentum delivered the virtual document which allowed organizations to create information once and reuse it many times across the enterprise. The company had produced three products: Documentum Server, Documentum Workspace (GUI Client), Documentum Toolkit. The target markets of Documentum are pharmaceutical, build to order manufacturing, regulated industries and government.

Overview of Venture Capital Company: Xerox Technology Ventures

The Xerox Corporation established Xerox Technology Ventures (XTV) in 1990 with the objective of commercializing innovations that otherwise would have been developed outside of the company. However, the venture firm had also developed projects that were not directly related to the business of Xerox. XTV had been relatively successful, and had invested in over a dozen companies in a variety of technologies: electronic publishing; document processing; electronic imaging; workstation and computer peripherals; software and office automation. The chairman of Xerox committed $30 million to XTV's fund, which was to be used to nurture and manage start-up ventures.

General Aspects of Venture Capital Firm

One of the main problems faced by any venture capital fund tied to a larger organization is a conflict of interests. Many corporate venture programs were formed with
the dual (and sometimes opposing) objectives of generating profits for its investors,
while promoting technological innovation within the corporation. Xerox structured
its fund in such a way in anticipation of this problem, meaning that XTV was
essentially organized as an independent venture capital fund. The fund's
management remained independent from the technological decision making in order
to be able to impartially evaluate the commercial potential of the technology. While
XTV was a corporate division, its legal contract with Xerox resembled the typical
agreement between limited and general partners. The main goal of the fund was to
maximize return on investments. It intended to eventually give up control of its
start-up companies.

Another way in which XTV resembled the independent venture capital organization
was in the spin-out process. A researcher would submit a preliminary business plan
to XTV, which would include a list of resources needed, such as laboratory
equipment, employees, or patents. The concept was then reviewed by XTV's board,
and if the technology passed the screening process, XTV would establish a start-up
company. One of the main challenges that XTV faced was establishing syndication
relationships with other venture capitalists. Syndication provided a valuable
additional screening process in addition to additional leverage. However, other
venture capitalists were reluctant to invest in XTV firms because the fund could not
make reciprocal investments in their funds. The charters with their respective parent
organizations limited the scope of the fund's investments. However, in the case of
Documentum, XTV was able to form a syndicate partnership with four other
venture capitalists. XTV's role was of incubation and follow-on rounds investor.

XTV structured its portfolio companies as separate legal entities with their own
boards and officers. The fund made a point of hiring management from outside of
the parent organization. XTV often recruited CEOs and managers from other start-

ups who were experienced in managing new enterprises. By hiring outside management, XTV hoped to avoid domination by Xerox's management and to keep the company focused on maximizing return on investment. The general partners of XTV played active roles in oversight and management of the portfolio firms as board members and informal advisors. The independent organization of these funds extended to the compensation structure for the fund's managers of the typical 20% carried interest.

XTV employed several policies that were unique to venture capital, which were intended to increase the risks and the incentives of the spin-out company's employees. First, XTV gave the company's employees options to buy shares of stock in the company. Second, the initial stake of the management team was small, and then increased with the company's progress. By the time of the IPO, the management might hold a 20% equity stake. Finally, unlike most internal corporate venture programs, employees who joined XTV spin-out firms were not guaranteed jobs at Xerox.

Relationship between the Venture Capitalist and the Portfolio Firm

For over a decade, Xerox had undertaken a large number of projects in object-oriented document management systems. XTV decided that this was a promising area, and recruited two engineers, Howard Shao and John Newton, who were executives at a relationship database manufacturing company, to head the technical effort. In the first six months, one of the engineers assessed the state of Xerox's knowledge in this area - including reviewing the several comprehensive business plans prepared for earlier proposed products - and assessing the market. He realized that while Xerox understood the nature of the technical problems, it had not grasped how to design a technologically appropriate solution. With the help of the XTV officials, Shao and Newton led an effort to rapidly convert Xerox's accumulated knowledge in this area into a marketable product. Xerox's accumulated knowledge gave Documentum a substantial lead over its rivals. This competitive advantage was enhanced by XTV's financial support of the firm during the Gulf War period, when the willingness of both independent venture capitalists and the public markets to fund new technology-based firms abruptly declined.

XTV and four other venture capital firms invested in Documentum.[112] The partnership closed $7 million of second-round financing in October, 1993 and $5 million third-round financing in September 1994. Documentum was now considered a later-stage investment as it had achieved profitability at the end of 1994.

XTV performed the role of traditional venture capitalists, which included screening proposed transactions, recruiting a management team, and attending board meetings of its portfolio firms. Also, because of its position as a corporate venture capital program, XTV allowed the spin-outs to leverage the substantial assets of Xerox. Probably the most important advantage for an XTV company was the certification provided by Xerox. The corporation allowed firms that it owned a majority of to be termed "Xerox companies" those that it owned 33 to 50 % of, "Xerox-alliance companies." This affiliation allowed XTV start-ups to sell to large companies that they would otherwise not have been able to. For instance, Documentum sold a system to Boeing for use in developing the service and customer documentation of its next-generation aircraft, the 777. Had Documentum been an independent start-up company, it would have been very unlikely that Boeing would have made such a critical purchase from them. Xerox implicitly guaranteed that it would stand behind Documentum if Boeing encountered any problems.

A spin-out firm like Documentum also had access to Xerox's world-wide network of suppliers, which often meant better prices and quality insurance. The firm could contract with Xerox to manufacture their products, at least initially, rather than making costly investments in its own facilities. Xerox also provided assistance in

---

[112] The other major investors are: Brentwood Associates, Merrill, Pickard, Anderson and Eyre, Norwest Venture Capital Management, Sequoia Capital.

the area of business services. Firms were expected initially to use Xerox's office space and equipment, as well as the corporation's accounting and legal services. XTV reimbursed Xerox at market rates. As the firm matured, however, it were free to employ other firms. A final, somewhat less important, source of assistance was Xerox's sales force, although XTV firms generally hired their own distributors. All of this meant that XTV had the advantage of access to Xerox's administrative and manufacturing resources, while it retained full autonomy in monitoring, exiting, or liquidating decisions. Documentum plans to go public in the first quarter of 1996.

4.2.4   Technology Transfer Process

Overview of Portfolio Company: Illinois Superconductor Corporation

Illinois Superconductor Corporation (ISC), located in Evanston, Illionis, was formed in October 1989. Its revenues, generated almost entirely from research and development contracts, totaled $202,348 for 1992. As of October, 1993, the company retained 14 full-time and 3 part-time employees. ISC developed and manufactured high-temperature superconductor components. These superconductors, when cryogenically cooled, enabled resistance-free transmission of electrical current. This unique property of the materials had tremendous commercial potential in the telecommunications, electrical utility, and defense communications industries. At the time of the IPO in 1993, Illinois Superconductor Corporation had licensed thirteen superconducting patents, or patent applications, from Argonne Laboratories and Northwestern University.

The firm had already been awarded one patent, and had four applications pending. ISC had used these discoveries to develop several promising technologies. These included a process which allowed high-temperature superconductors to be applied to surfaces through a process similar to painting. This was a cheaper alternative to the currently used method. A second innovation was a superconducting sensor that provides continuous readings of the temperature of ultra-cold refrigerators that store human tissue. Technicians monitoring these refrigerators had previously been required to periodically open the units to check the levels of refrigerant. ISC's sensor was one of the first high-temperature superconductor devices to be commercialized. Ultimately, the firm hoped to develop a variety of signal processing and filtering components for the cellular telephone and wireless communication industry that would employ superconducting materials.

Overview of Venture Capital Company: ARCH Venture Partners

ARCH Venture Partners was established in 1986 to commercialize promising research from the University of Chicago and Argonne National Laboratory in what is known as the technology transfer process. The technology transfer process generates revenue in one of two ways, either through licensing the technology to firms, or through establishing spin-out companies. ARCH was created in order to generate revenue, and to contribute to the regional economy by creating new enterprises and jobs. ARCH Development Corporation was created as a separate, private, not-for-profit corporation in affiliation with Argonne and the University of Chicago.

As of October 1993, ARCH Fund's first fund, with a total $9 million, had invested in twelve firms. The limited partners included the University of Chicago, State Farm Insurance and two venture firms. ARCH Development Corporation was the only general partner. ARCH Venture Partners II, L.P. was intended to raise $30 million. The fund specialized in new and early-stage technology (primarily biotechnology) companies located in the north-central mid-west region of the United States.

General Aspects of Venture Capital Management Strategy

The first stage of the technology transfer process took place in the Argonne National Laboratory and in the laboratories of the University of Chicago. Scientists were required to report any "useful" discoveries. ARCH reviewed these preliminary reports, and if the discovery appeared to have commercial promise, ARCH claimed title to the invention. ARCH then decided how to commercialize the technology by reviewing the trade press, consulting academic and industry experts, and conducting limited surveys of potential customers. In evaluating an investment, the fund considered the following issues : the extent to which the discovery could be protected through patents or other means; the difficulties that would be encountered in scaling up the manufacture of the product or process; the potential for market acceptance and the extent of likely competition; and the likelihood of rapid obsolescence. After making these tests, if the product appeared to be promising, ARCH initiated the process of seeking intellectual property protection.

Once ARCH identified a technology that would potentially form the basis of a new business, the fund provided a minimal amount of early seed financing to develop the technology and a business plan. The plan was presented to the ARCH executive committee (which included both ARCH officials and members of its board) for a review, and if approved, ARCH contacted other venture capitalists to propose a possible co-investment. ARCH managers also encouraged the start-up company to appeal to non-equity sources of funds, such as the U.S. Government's Small Innovation Research Program. ARCH staff members occasionally served as temporary general mangers, but attempted to replace themselves with recruited entrepreneurs as soon as possible.

Relationship between the Venture Capitalist and the Portfolio Firm

ARCH Venture Fund I was the lead seed investor in Illinois Superconductor and led two additional venture investment rounds totaling $3.2 million. The financing of this firm was typical of ARCH's firms. In October 1989, in conjunction with the formation of the firm, the company issued 136,000 shares of common stock to ARCH. In its Series A financing (undertaken during 1990 and 1991) the company raised a total of $1.5 million, from ARCH Venture Partners, and another Chicago-based venture capital organization, and the Illinois Department of Commerce and Community Affairs. Each investor contributed $500,000. The firm raised several more million dollars from these same sources and others in two additional rounds in 1992.

The connection to Argonne Laboratories and ARCH Partners gave Illinois Superconductor an important advantage in the high-temperature superconductor market. ISC licensed six inventions for basic superconducting processes from Argonne, and if its patents were granted, ISC could earn licensing fees from other companies that use the technology. The company also was given a priority on any discoveries Argonne made in its industry. This gave the company a particular advantage over the competition.

Another advantage for ISC was the relationship to ARCH. Steve Lazarus was recruited as the head of ARCH. He acted as a director of ISC from January 1992 until August of 1993, when he became the company's Chairman of the Board of Directors. One of the advantages for an upstart firm, such as Illinois Superconductor, in a relationship with ARCH was Lazarus' networking ability. Lazarus' partners, ARCH associates (typically with former employers), former associates, and ARCH's board provided an informal network. Such connections enabled Illinois Superconductor to raise capital from other venture capitalists.

Many in the industry were reluctant to consider seed investing due to the large amount of time and effort needed to monitor a seed investment. Also, venture capitalists based in other areas of the country were often reluctant to invest in firms based in the Midwest because they wanted to provide close oversight of their early-stage investments. Finally, ARCH was limited by its charter to investing in deals where the technology had originated at the University of Chicago or Argonne National Laboratory. Consequently, ARCH could not invest in most transactions initiated by other venture organizations. Since reciprocity was an important aspect of the venture investment process, the prohibition had limited ARCH's ability to build strong ties with the venture community. Therefore, Lazarus was forced to aggressively seek out relationships with the venture community. He sought to exploit old connections and new contracts from the University of Chicago.

ARCH's role in ISC resembled that of a typical venture capital fund: consulting informally with management, attending board meetings, and making decisions about whether the firm should receive second- and later-round financing. If the company's venture backers decide to terminate the firm, the technology reverted back to ARCH, who in most cases, sought to license it. One of the primary goals of the ARCH partners was to add value through the provision of oversight. Therefore, like Illinois Superconductor, most of its spin-out firms were based near Chicago. The company had gone public in 1993 and had been ARCH's most successful spin-out to date.

4.2.5   New Directions in Venture Capital : Venture Leasing

Overview of Portfolio Company: RhoMed

RhoMed is a biotechnology firm specializing in developing radio-pharmaceutical (i.e. nuclear medicine) products for diagnosing and ultimately treating diseases. The company had developed a method to label antibodies with small amounts of radioactive element, Technetium. In theory, these cells could be injected into the human body and tracked to identify a variety of potentially serious conditions. The company was established in 1986. RhoMed's founder was Dr. Buck A. Rhodes, who also served as RhoMed's president and chief scientist. He had begun working as a researcher in the 1970s at John Hopkins University in Baltimore and changed in 1978 to University of New Mexico, close to Los Alamos National Laboratory. With an increasing interest in business, Rhodes joined a local biotechnology concern, Summa Medical Corporation, as the senior vice president for scientific affairs. In 1986, Rhodes left to begin a firm of his own. Beginning the firm out of his garage, Rhodes awarded funds from the Small Business Innovation Research program within its first two years of operation. In addition to developing Rhodes technology,

RhoMed signed Cooperative Research and Development Agreements (CRADAs) with Brookhaven National Laboratory and Los Alamos. The company's revenues, from grants and contracts, sales, and license fees and royalties, totaled $760,622 for 1992.

RhoMed had received most of its initial financing from three sources: government grants designed to promote private sector research; collaborative agreements with national laboratories; and strategic alliances with large pharmaceutical firms. In the venture lease deal with Aberlyn, RhoMed would receive $1 million in capital along with the option to purchase its patent back for one dollar at the end of the three-year lease.

Overview of Venture Capital Company: Aberlyn Capital Management

Aberlyn was established in 1989, as an investment banking firm for the biotechnology and biomedical industries, providing oversight in mergers and overall business development in addition to capital. In 1992, the firm expanded its services to include venture leasing. With its venture leasing business growing in importance, Aberlyn established Aberlyn Capital Management Company (ACMC) as a subsidiary. Under ACMC were two other organizations: BioQuest Venture Leasing Company, N.V., which served to finance the lease arrangements, and Aberlyn Capital Management Limited Partnership, which acted as the general manager of the leasing fund. As mentioned above, one of the perceived problems with venture leasing was the extensive administrative requirements. ACMC contracted out the billing, data processing and financial reporting tasks to another leasing firm. In this way, ACMC maintained a lean organization which allowed its management to focus on generating new business. An important aspect of Aberlyn's structure was its partnership with a large Dutch investment bank, MeesPierson, N.V., which provided operating capital, and played a key role in raising funds from individual investors in Europe and the Middle East for Aberlyn's leasing fund.

In 1992, Aberlyn's management team had developed the concept of the "FLIP" -- Finance Lease on Intellectual Property, which provided leases based on patents. This would allow growing firms to finance their need for working capital through leasing, rather than just equipment purchases. Aberlyn would purchase one of the firm's patents, and the firm would then lease the patent from Aberlyn and therefore gain the legal right to use the patent. Aberlyn would hold or own the title to the title to the patent until the lease had expired, at which point the lessee could exercise its option to purchase the patent at a nominal price.

General Aspects of Venture Capital Management Strategy

The case of Aberlyn Capital Management and RhoMed takes the innovative venture leasing business one step further from the leasing of tangible assets to the leasing of intellectual property. However, the same issues concerning valuation, compensation, risks and returns of the venture leasing industry apply. Venture leasing is essentially a hybrid between venture capital and banking, and reflects the increasing diversification within the private equity industry. The first venture leasing firms began in the late 1960s when the venture capital market declined and venture capital firms tried to find new markets for their services. With venture-leasing, venture capitalists lease equipment to new firms in addition to providing equity financing. For the start-up firm, in addition to often being the only feasible financing option available, venture leasing provided several advantages over secured debt. Banks and traditional leasing agencies were often reluctant to deal with start-up firms without imposing severe costs and restrictions.

Despite several successful transactions in the early 1970s, the industry met with several problems which led to its decline. Start-up firms were wary that leases would provide the venture capitalists with too much control over the firm. Venture capital firms were also reluctant to undertake management of leases due to the great amount of administration and oversight involved. Finally, the general economic conditions of the 1970s led to a reduction in the capital commitments to the venture capital industry. Therefore, venture capital funds concentrated their efforts on raising funds for new firms. Venture leasing rebounded in the early 1980s due to the growth of the semiconductor industry, which required expensive manufacturing equipment. The success of these early venture leasing firms served as a catalyst to sustained growth of the industry.

In general, the venture lease contract follows the typical venture capital model, with the funds structured as partnerships in which the venture lessors acting as the general partners. Until recently, individuals have primarily been the limited partners, although foreign individuals and institutions have increasingly entered the industry. Institutions, such as pension funds, have reluctant tot invest in these funds due to the high information costs and risks involved. Another concern is the tax status of distributions for non-profit institutions.

In terms of the lessee, the venture leasing company has three sources of compensation: lease payments; purchase options at the end of the lease period; and warrants. The fund can adjust these three in order to create a more bond-like or more stock-like return. One of the main differences from venture capital funds is that investors may have to wait years (for an acquisition or IPO) until they realize any returns. Venture leasing funds, on the other hand, generate returns almost immediately through lease payments, and pay out distributions on a quarterly basis. The regular lease payment typically reflects a spread of 1 to 10 % over a Treasury or prime base interest rate, with the spread being a function of the risk of the investment and the other financing and investment options for the involved parties. For example, the scarcity of available venture capital financing and the decline in the IPO market resulted in an historically high spread in the summer of 1993. The lessee also pays a certain amount to the venture company in order to reserve the right to purchase the equipment at the end of the lease period. Finally, perhaps the most potentially rewarding aspect of the compensation structure are warrants. These warrants represent a right to purchase shares of the company's stock.

As a relatively new asset, the venture lease posed several risks to Aberlyn in addition to those inherent in a traditional venture capital investment. Aberlyn felt that one of the most important ways to limit its risk was to keep a number of technical experts on its staff and its advisory boards. Being able to evaluate promising technology would also allow Aberlyn to extend its leases to new firms with potentially highly profitable innovations. Most of its competitors limited their lease financing to cover laboratory equipment and computers. RhoMed's patent exemplifies Aberlyn's interest in widening the scope of its leasing business in order to take advantage of emerging technologies.

Aberlyn employed other safeguard measures which were common to the venture leasing industry. The first concern was protection from ex ante risks, that is risks inherent to the potential lessee. Aberlyn made it a policy to only invest in firms that had received at least two rounds of venture financing, in order to ensure that the firm had sufficient cash flows or reserves to meet the lease payments. The lessee firm was then ranked on a scale of 1 to 5 (with 5 being the lowest risk) on the basis of the likelihood that it will be able to make the lease payments. Aberlyn's policy strategy was to diversify its lease portfolio across these ranks (5% in Class 5; 15% in Class 4; 35% each in Classes 3 and 2; and 10% in Class 1). It was also important for Aberlyn to protect itself from risks that occur after the agreement is signed. This was accomplished through the specific terms written into the contract. Aberlyn sometimes required that a nonvoting representative be placed on the firm's board of directors. (A voting director would mean that Aberlyn might jeopardize its legal status as a lender in the case of a bankruptcy).

Relationship between the Venture Capitalist and the Portfolio Firm

The venture lease arrangement was advantageous from RhoMed's point of view for several reasons. The company's goal was to assure the firm's progress while maximizing their control over the firm's equity. First, this method of financing would entail a minimal reduction in their ownership stake. Second, RhoMed saw that Aberlyn had the ability to raise capital from European and Middle Eastern sources. This leasing transaction could serve as a bridge until a private placement. The private placement would then assure financial stability until an IPO. The CFO of RhoMed thought that an appropriate time for an IPO might be in October 1994, after its product development effort was further advanced.

The patent on which the lease was based was valued at nearly $5 million by an outside consultant. In the proposed transaction, RhoMed would receive $1 million from Aberlyn, and Aberlyn would be granted a seat on RhoMed's board. However, Aberlyn would most likely play a relatively passive role of limited engagement with RhoMed. Typically, the venture lessor leaves the role of monitoring to other follow on investors who provide more substantial financing. The lease actually made for $1.15 million, but Aberlyn would retain $150,000 to cover the first year's interest, and RhoMed would not have to make any interest payments until the beginning of the second year. RhoMed could purchase its patent back for one dollar at the end of the three-year lease after repaying Aberlyn's loan. The interest rate on the outstanding amount of the lease would be 15%, as the firm was classified as Risk Class 2, and a warrant coverage of 10%, with exercise price at $1.45. (Therefore, Aberlyn would receive approximately 80,000 warrants).

4.3 The German Studies

4.3.1 Successful Trade Sale

Overview of Portfolio Company: EOS GmbH Electro Optical Systems

EOS is a supplier of laser optical equipment and the relevant computer periphery (computer graphics and CAD and so on). The firm was founded as a marketing organisation in April 1989. At that time EOS developed two innovations: one was the EOScan, a three-dimensional system for the measuring of coordinates for the fast, contact-free identification of surface coordinates of three-dimensional objects; the other was stereolithography for the fast creation of complex models from CAD data without the use of tools (called STEREOS). Both were financed in 1990 with venture capital. These two products have to be seen in conjunction; they form the concept of 'rapid prototyping'. Thanks to a DM 700,000 development contract from the automobile manufacturer BMW, the development work for the stereolithography system was able to proceed quickly; in 1991 the first prototype for STEREOS was ready. As well as these two families of products, EOS also supplies an article of merchandise. This is a laser interferometer for the measurement of planarities of surfaces (Flat-Master), for which EOS has the exclusive selling rights. In 1993, another product line for laser-sintering was introduced (EOSint).

The competitive advantage of EOS lies in the fact that the systems rely on leading technology and are market oriented. Their machines are faster, more productive and easier to handle than those of their competitors. At the same time, EOS offers a high degree of quality in the accompanying technical services and in the qualification of its employees. Right from the beginning, EOS targeted the following groups: the German automobile industry and the electrical industry (e.g. Siemens and

Electrolux household equipment). To these were added medical technology and casting technology.

Overview of Venture Capital Company:Technologieholding

The Technologieholding was founded in 1987. Due to the relatively long time needed for fund raising, business activities were only able to commence in April 1990, the Technologieholding concluded its first two investments, with its first fund , the ETH.

The problems in acquiring capital were the following:

- the subject of new technology based firms (NTBFs) is no longer fashionable with investors
- venture capital companies in Germany have not been very successful so far
- the Technologieholding had no previous history or evidence of investment returns.

The first fund of the Technologieholding, the ETH, started up with 6 million Gulder (just under DM 6 million, 1990). At the end of 1995 the volume invested was just under DM 100 million (45 portfolio enterprise, all of them new technology-based firms). The investors in the ETH fund are: international investors from US, France, the Netherlands, Finland, Switzerland, Austria and Germany.

General Aspects of Venture Capital Management Strategy

The business purpose of the Technologieholding is high return on investment. The aim is to achieve, together with the entrepreneur, the greatest possible increase in the value of a firm over the period of the investment. The European Technologieholding N.V. (ETH) has its head office in Amsterdam and is managed by the Technologyholding VC GmbH, Munich.

The ETH fund invests venture capital in young, innovative enterprises in their early stages (foundation, market introduction, expansion) in all areas of industrial electronics (e.g. sensorics, computers, computer periphery, software, automation, control and measurement technology, laser optics and electrooptics, communication technology, process control and image processing). They concentrate on markets with a high growth potential, and on product and market areas in which it possesses know-how and experience. It is definitely prepared to invest in high-risk enterprises; however, these must evidence outstanding chances of success. Enterprises should preferably be situated in Germany, France, Austria or Switzerland. These areas account for about three-quarters of their investments. The share in the enterprise is usually between 25 and 50 %.

Before making an investment, the Technologieholding conducts an intensive survey and evaluation of the enterprise applying for capital. This investment survey is carried out by the management company itself and requires about 10 to 50 man days in each case. The Board of Directors of the fund comprising of internationally experienced managers is finally responsible for the investment decision.

Important evaluation criteria are:
- Technology area and growth potential of the firm must correspond to Technologieholding's investment policy,
- Competitive advantage of product,
- Size of market targeted,
- Quality of management,
- Value added potential.

Technologieholding pursues the strategy of "hands-on" management (very intensive consulting, on average 0.5 to 2.0 man days per month per investment portfolio), with the consulting services being paid for by the portfolio enterprises. The Technologieholding provides back-up for its portfolio firms in the following areas:

strategies, business management, financing, marketing, contract design, crisis management.

The Relationship between the Venture Capitalist and the Portfolio Firm

EOS was able to finance the first business year through the BMW development contract. Since there was a further need for capital, the enterprise actively tried to attract investments right from the start (both for the R&D phase and for market introduction).

Based on negotitations with public equity providers, the negotiations with the Technologieholding, with whom personal contacts existed, were long drawn out and led to temporary liquidity shortages. At the end of December 1990, the investment contract was concluded and Technologieholding (together with the Tbg, under the BJTU pilot scheme) invested DM 1.8 million. In summer 1992 Mr. Langer called an option through which he was able to increase his share by 75%, although he had already at this point signed the share contract with the Technologieholding. In addition, since the enterprise's foundation, it has benefited from various promotion funds, receiving DM 738,000 from the Bavarian innovation promotion programme BAYTEP, DM 310,000 from the R&D project for the EU programme BRIT-E/EURAM, under its EKH-programe (loans) of DM 112,000 from the DtA (1989).

Technologieholding advised EOS and provided back-up services in entrepreneurial matters during the whole investment period. To begin with, this took place within a separately concluded consulting contract, which was terminated after two years but advice continued to be given as before.

The back-up services consisted primarily of strategic advice and management support. Technologieholding was particularly helpful in all financing questions; Technologieholding also supported the development of a controlling system. EOS described the support as very useful and expert, and the relationship to

Technologieholding as one of partnership. The investment in EOS was harvested by Technologieholding by a trade sale.

The sales of EOScan and STEREOS showed a continuous upward trend. In the business year 1992/93 a turnover of DM 8.5 million was reached. At that time EOS had 25 permanent employees. In December 1993 the Technologieholding sold its share in EOS to Zeiss and was thus able to realize an investment return of 75 % p.a. (according to the EVCA guidelines). The sale took place without any particular problems. Zeiss was interested in EOS particularly because of the technology.

4.3.2   Transatlantic Investment

Overview of Portfolio Company: Tomtec Tomographic Technologies GmbH

The firm of Tomtec Tomographic Technologies GmbH was founded in June 1990. One prerequisite for this was the confirmation of a large development contract from Siemens. At that time the main product, which also constituted the product idea for the foundation, was a so-called 3-D-Echo Computer Tomograph. This was a medical technology product which could be connected up to existing ultrasonic devices ("add-on product"). It enabled three-dimensional images to be obtained (ultrasonic scanners otherwise give a 2-D-image), which are much more suitable for diagnostics.

The market targeted for the 3-D-Echo-CT comprised university clinics using ultrasonics in cardiology, world-wide. It was also intended to sell the product to OEM-customers (ultrasonics producers). One year later the product was to be sold to the diagnostic sector, i.e. normal hospitals, and later on to doctors' practices. The other product lines are for the same market; however, they are already being sold to practices and hospitals. A first prototype of this "real" innovation (a world first) was completed by November 1991. The R&D expenses (including external expenditure) up to that point amounted to between DM 5 and 6 million, the time taken being

about 2.5 years. At this point market introduction took place. Despite a positive
reaction and many interested parties, sales were only realized from the end of 1993
on.

In 1993 a turnover of DM 2.5 million was achieved with the 3-D-Echo-CT, 80 %
of this in the last quarter of 1993. In 1995, Tomtec had a total turnover of DM 16.6
million, 118 employees and returns of DM 2.2 million (all figures provisional). A
great part of the turnover was bought in by the acquisition of a firm in the US which
also makes ultrasonic products. Tomtec's range now consists of: the 3-D-Echo-CT,
networks for ultrasonic cardiology applications, Stress-Echo (a stress
electrocardiograph with ultrasonics). The Tomtec partners include six VC
companies with a 75% interest altogether (Allstate, Atlas Venture, Frontenec,
Marquette, MVP, Oxford). The remaining shares are held by the management and
the founder, who acts as adviser responsible for strategic product development and
new business fields.

Overview of Venture Capital Company: Atlas Venture

The Atlas Venture group, with offices in Amsterdam, Boston, Paris and Munich,
was originally founded in 1980 as an investment company of the NMB Bank
(Netherlands Mittelstand Bank, now the ING Bank). This means that Atlas Venture
developed out of venture capital activities undertaken by the NMB Bank in 1980. In
a buy-out in 1986 the management went independent and began to acquire other
investors. Atlas has been active as an independent investor since 1987. Its partners
now include the ING Bank, PPGM, Centraal Beheer and several Dutch pension
funds. Atlas Venture is a venture capital company which pursues the aim of
establishing itself as one of the important internationally-active participants on the
VC market.

General Aspects of Venture Capital Management Strategy

Today the group has 40 employees, whom of half are working in investment management. Atlas Venture manages a financing volume of DM 400 million. The 80 portfolio enterprises - 18 of them in Germany, Austria and Switzerland - cover a wide range of financing situations, from start-ups, through MBO financing to re-structuring. About US$ 50 million are newly invested every year, one-quarter of them in German-speaking countries. Technology financing is a dominant field for Atlas (USA 100 %, Netherlands 50 %, Germany 75 % of portfolio enterprises). Atlas Venture concentrates on two main areas:

- Venture capital for young, innovative firms in the areas of information technology and life sciences (biotechnology, pharmaceutics, medical technology, environmental technology) with chances for international expansion in the early development phases. The turnover aimed at within the planning period should be c. DM 30-35 million;

- Investment capital for "Mittelstand" enterprises (small and medium-sized firms) with growth potential in Germany. Here, Atlas Venture invests without particular sectoral or technological emphasis in established enterprises seeking a partner with a long term orientation for the financing of a change of ownership, turn-around or international expansion.

Atlas Venture invests sums of between DM 1 and 5 million; if larger amounts are needed, the group has the experience to assemble and coordinate consortia of several venture capital companies. It aims at a minority interest with long-term active and strategic collaboration on advisory boards, boards of directors and other committees.

Atlas Venture's support for portfolio enterprises is mainly in the form of hands-on management, without interference in business operations. Rather, it regards its task as bringing together a suitable management team and providing support on strategic issues. In crises, Atlas participates in decisions to counter crisis (e.g. bringing in an external crisis manager, appointing new management, developing a new strategic concept). The prerequisites for investment are competent, experienced management, competitive advantage (technological or other), a good market position, strong growth potential. Atlas carries out a very stringent pre-investment survey, investigating all relevant aspects of the enterprise (management, product, market, etc.). The pre-investment survey is based on the business plan, own estimates and information from personal discussions, analysis of market studies, specialist publications, sectoral analyses, data bank searches, questioning of experts, interviews with potential customers, etc. As Atlas Venture has branch-specific and technology-specific knowledge, technical expertise is only commissioned externally in exceptional cases. However, reference discussions are held with experts.

The Relationship between the Venture Capitalist and the Portfolio Firm

The first seven months of business activity (and development work) of Tomtec were funded by the development contract from Siemens (DM 1 million). Following this, additional financing of just under DM 4 million was acquired in 1991 (DM 1 million from Atlas Venture, DM 1 million from the Tbg through the BJTU pilot scheme, DM 1.8 million from the BAYTEP programme (Bavarian technology programme, long-term loan). Apart from Atlas Venture, Tomtec was also negotiating at that time with other venture capital companies, but none of the discussions resulted in agreements, as these companies were unable to judge the sales chances and were doubtful about the prospects of expensive medical products.

The capital acquired was very quickly used up in development work, so that an increase of capital from Atlas and Mikron took place. At the beginning of 1993 a new VC partner was sought, at first in Europe, but despite strong support from Atlas and Mikron no European partner could be found. At last a US American venture capital company was found, which invested US$ 5.4 million in August 1993. Following this capital increase the registered office of the firm was transferred from Germany to the US. Three months later a further capital increase of nearly US$ 10 million took place, in order to acquire the American firm mentioned above.

The greatest difficulty in the search for sources of finance was a lack of preparedness to take risks on the part of European investment companies and banks, which were not willing to commit themselves to any long-term financing. Thus it was not possible for Tomtec to acquire European venture capital to finance further growth, despite massive support from Atlas Venture.

The provision of capital by VC companies is linked with free support and advice. Four times a year an official "board meeting" takes place. In addition, an "executive meeting" is held six times a year. Each of the seven VC companies that now have shares in Tomtec holds a meeting once a year, at which portfolio enterprises introduce themselves and these enterprises are discussed. There are also informal telephone, written or personal contacts at least once a week, with the initiative coming from both sides.

Assistance is mainly in the strategic area of business management, in which VC companies possess a high degree of competence. Financing is also always a main point of discussion at board meetings. The VC companies also procure qualified personnel and mediate contacts to cooperation partners, banks, consultants, lawyers, tax advisers, investors, potential clients and other enterprises. Tomtec assesses this assistance and advice as being very useful. All the VC companies on the „board" are very professional, none of them interferes with the day-to-day running of the firm, they all have a lot of contacts. Since several VC companies have an interest, more opinions, more competence, more contacts etc. can be brought in altogether. Up to now it has not been possible to identify a decision about means of exit.

4.3.3   Regional Investment

<u>Overview of Portfolio Company: BioTec[113]</u>

BioTec was founded in December 1986, with the reason that the founder wished to go independent. The field of activity of BioTec is subdivided into two areas:

-    Investigating ground for toxic substances.

-    Detoxification: biological ground treatment. For this purpose an "in-situ" detoxification treatment has been developed, in which the ground is cleaned by a special process involving microorganisms. This biodegradation process enables the treatment of toxic substances to take place far below the surface (chlorohydrocarbons), which is not possible with the method of removing contaminated ground. BioTec develops the detoxification concepts, methods of treatment and biodegradability studies. The treatment itself is then not carried out by the NTBF but by a third party. The development expenditure for this process was DM 1.2 million.

The markets targeted by BioTec in the identification and detoxification of toxic deposits are petrol stations, the chemical industry, the metal processing industries, dry cleaners, refineries, emergencies caused by accidents, waste deposits (reconnaissance only) and the building trade. The market introduction of the detoxification process took place early in 1989 shortly before the TOU promotion support ran out. To begin with, the turnover developed very well but did not cover the costs.

<u>Overview of Venture Capital Company: Mittelständische Beteiligungs-gesellschaft Hessen</u>

The MBG (Mittelständische Beteiligungsgesellschaft Hessen) was founded in 1971. The founder partners were the Bankverband Hessen e.V., the Hessische Landesbank

---

[113] Name of the company is changed.

- Girozentrale -, and the the Genossische Kreditinstitut (cooperative central savings banks of Hesse and the Hessische Landesentwicklungs- und Treuhandgesellschaft mbH (HLT), Hesse's association for the promotion of regional industry, which, since the foundation, has undertaken business on behalf of the MBG.

General Aspects of Venture Capital Management Strategy

The aim of the business at that time was fixed as "the acquisition and administration of investments in enterprises in small and medium-sized firms, in accordance with the federal policy of supporting the investment financing of small and medium-sized firms". In the first business phase of the MBG Hessen, 108 investments with a total volume of DM 20 million were granted up to 1981. This was accompanied by successive capital increases to DM 1.3 million and the participation of the Handwerkskammer Hessen (Hesse Chamber of Handicrafts) as a further partner. In 1978, to complement the normal business stipulations, a firm foundation programme was initiated specifically to assist new foundations.

In 1984, the capital of the company was increased to more than DM 2 million and at the same time the Chamber of Industry and Commerce as well as the employers' associations were acquired as new partners. Thus the whole of industry in Hesse had acquired an interest in the MBG. This was also reflected in a re-formulation of the business policy regarding the statutory tasks of the partners: a strictly subsidiary orientation; initiation financing of projects with good success prospects; investment in small and medium-sized firms with good profit expectations, original (innovative) products and high value-added, situated in Hesse. On this basis, the MBG made a total of 235 investments in small and medium-sized firms in Hesse up to the end of 1994, with a volume of c. DM 89 million. These have included 101 innovative firms (volume over DM 50 million). The failure rate is about 20 percent. Current investments at the end of 1994 comprised 80 enterprises (total volume just under DM 42,5 million).

The MBG Hessen only allocates silent partnerships (over 10 years). The lower limit for investments tends to be about DM 250 thousand, the upper limit about DM 2 million. Return on investments consists of a fixed basic amount and a profit-dependent arrangement (in standard investments, 7.5 % fixed rate p.a., plus 0.7 % guarantee commission, plus two percent dependent on profits).

In principle, the MBG Hessen is prepared to co-invest with all commercial investment companies, but these are very cautious regarding the MBG's target group, as they are not willing to take risks.

To generate a "deal flow", MGB Hessen practices active acquisition through public relations and events, e.g. for credit institutions. Generally, MBG Hessen pursues a rigorous evaluation and selection process. This involves intensive use of external advisers and external institutions as well as the network of the MBG Hessen.
The evaluation criteria are:

- Business plan: most important is the intensive examination and discussion of firm concepts with respect to plausibility, among other factors.
- Business situation: there should not be any acute liquidity shortages. All available information (e.g. on accounts, current evaluations, contract situation, individual calculations of business contracts) are subjected to a detailed examination. MBG talks to the company's banks and brings in tax advisers.
- Product and market: the product should be superior to similar products, either through a technology lead or due to its level of innovation. In order to evaluate the product and the market, an expertise is first performed, e.g. by external independent institutes, paid for by the enterprise. With innovative products, the enterprise itself has to provide data on markets (e.g. market volume). The MBG Hessen also makes use of its network and asks other portfolio enterprises.
- Firm founders: MBG wants to know their specialist background and previous activities, evaluates their entrepreneurial stability, ability to

cooperate, ability to learn. Since "ready-made" businessmen do not exist, the assessment of development potential and ability to learn is one of the most important criteria. This relates to the fact that although the MBG provides advisory services, these are not very comprehensive, due to the MBG's promotion aims and to capacity and cost restrictions. Judging the founder's personality is difficult and is also difficult to describe. It cannot be operationalised and is thus partly a subjective decision.

The portfolio services, which could be characterized as "hands-off", are not paid for by the recipient. The yearly expenditure of MBG Hessen on support services amounts to about 5 man days per year per firm. Each of the MBG's portfolio supervisors is responsible for servicing approximately 15 enterprises, visiting each about twice a year. Other servicing activities ("network" activities) take the form of an informative newsletter circulated three to four times a year, an annual management seminar and introducing the portfolio enterprises to one another, and awarding of an innovation prize by the MBG Hessen every two years.

The portfolio servicing activities concern the development of business strategies, controlling, accounting and planning, as well as crisis management. However, regarding crisis management, only analyses of weaknesses are carried out as a rule, these include turnover- and liquidity analysis and examination of the firm's business concept. In some cases technical specialists, tax advisers, business consultants, chatered accountants are brought in. In crisis situations, the MBG Hessen assists its enterprises in all crisis management contacts (e.g. interviews with banks).

The Relationship between the Venture Capitalist and the Portfolio Firm

BioTec had already received support under the TOU pilot scheme (subsidies of just under DM 1 million) with the necessary own share (DM 283,000) being financed by a loan from the Sparkasse Langen (Langen savings bank).

Since in the market introduction phase, which lasted about two years, no income could be made, a financing bottleneck occurred in 1989 when the TOU support ran out, so that almost the whole team of 8 employees had to be dismissed. The situation only changed in December 1991 through the participation of MBG Hessen. The investment negotiations began with a preliminary interview. Then the founder submitted an entrepreneurial concept, together with the statement of the KfA from the TOU pilot scheme. By the end of this preliminary investigation stage these factors served as a basis for the establishment of share participation. A report was drawn up and presented to the Commission for MBGs.

The founder stated that it was not the enterprise but himself that was investigated, e.g. how realistic his aims were, what personal impression he made, how convincing his explanations were. The MBG was not in a position to assess the technical issues. All in all, the investigation was relatively intensive, the employees of the MBG were competent. However, the banks made a more stringent investigation than the MBG Hessen.

Assistance was not given regularly by the MBG, but on demand. Otherwise, the MBG made a three-monthly visit to BioTec. The assistance is, or has been, concentrated in the area of strategy and financing (at present: negotiations on shares, marketing (here the advice is not in-depth), use of personnel, controlling/accounting). At the moment the MBG Hessen supports the introduction of a controlling system for commission- and project management. BioTec describes the servicing relationship as an intensive, cooperative partnership providing specialist skills. The usefulness of the individual discussions is rated very high; however, the three-monthly interval between visits is considered too long.

The investment share of the MGB Hessen takes the form of a silent partnership with a term of ten years. The partnership is bought back by the company on expiry of this term. However, the partnership can also be transformed into a long-term loan. On conclusion of the case study no decision had been made in this respect; but experience suggests that a buy-back is the most likely solution.

4.3.4 Early Stage Investments in Industrial Spin-Out and Scientific Spin-Off

Overview of Portfolio Companies:

MIOS

The company is a spin-out of a medium-sized software company named Logware. A manager founded MIOS in 1990 and took with him seven employees from his former company. MIOS works in the sector of the process automation of storekeeping and logistics. The company has now completed its development stage and, by and large, entered the standardization phase. It supplies standard products and systems for wireless data-communication including infra-red based data terminals and hand-held steering equipment on the world-market. Only a few companies are able to compete with the system solutions provided by MIOS. The production of the hardware takes place outside the company. MIOS is developing a network of hardware suppliers. The major part of the work now involves the development of systems solutions and adapting them to the customers' requirements. The customers mainly consist of companies supplying large equipment for storekeeping firms as well as big production companies in the automobile sector. Its annual turnover has grown rapidly to DM 3.5 million since it first started six years ago. Today the company employs 18 people, as compared to only seven at the beginning. The majority of them are software engineers. When it was founded, MIOS was subsidized by public support programmes which also contributed towards its capital. MIOS now obtains growing revenues from its products, whilst public support only plays a minor and continually decreasing role in its development.

Graphikon

The company was founded in 1990, as a spin-off of an institute within the Akademie der Wissenschaften of the GDR (ADW, Academy of Sciences). The reunification process did not offer the institute any existence in the future. Consequently, the head of the image processing department took on some of his former employees and founded Graphikon together with a business friend from a West German company. Graphikon operates in the sector of software and automation. After it was founded the company built up soon a staff of thirty employees predominantly from the ADW. In the old days of the GDR, the ADW department maintained contacts to large West German industrial companies, facilitated by its unique knowledge and attractive performance. Thus, after the spin-off was established, the company was still able to participate in large projects with West German companies. However, those contacts have not provided Graphikon with a stable financial base for development.

Today the company has two profiles: the first involves the large VECTORY system, a software solution for the automated digital recording of documents. The second and more important profile represents a more development-oriented company with three additional company sectors in the area of image processing. There are plans to develop product lines during the next few years and to transform the company sectors into profit centers. This strategy is being accelerated by a new business executive at Graphikon, an experienced business consultant from West Berlin. The original West German colleague left Graphikon in 1994. As far as the rest of the staff are concerned, the majority are software engineers. VECTORY, the only product which the company has introduced onto the market, is very development-based and service-oriented.

In 1991 an opportunity was offered by an Italian company (GEPIN S.p.a) to distribute its highly recommended products in Germany and to participate by supplying external know how. Consequently, a joint venture, Graphikon Systems AG, was founded in 1991. The capital share distribution was divided almost equally between the Italian (51%) and the German partners (49%). However, the Italian partner became increasingly insolvent, and the joint venture company was liquidated in 1994. The employees of the AG were taken on by the German parent company. The risky experience with Italian partners has induced the company to proceed with caution. After five years of existence growth has been moderate. The company has 34 employees and a current turnover of DM 3.7 million.

Overview of Venture Capital Company: The LBB-Seed Capital Fund

The idea of establishing a seed capital fund was based on the BJTU government programme, which helped to create a financial infrastructure for NTBFs. The fund was developed in 1990 as a joint venture between two Berlin institutions: VDI/VDE Informationstechnik GmbH (a technology transfer agency and administrator of public R&D programmes focusing on information technology and with extensive experience in assisting NTBFs) and the Industriekreditbank, a medium-sized bank that concentrates on financing industrial investments in Berlin. In 1994, the Seed Capital Fund (SCF) was sold to the Landesbank Berlin (LBB), a holding which is the largest bank company in Berlin today, and which was looking for investment opportunities to cover the segment of financing NTBFs, especially early-stage and first-rate expansion. The SCF is now managed by two businessmen, one from the Landesbank and one from VDI/VDE-Informationstechnik GmbH, the initiator of the fund.

General Aspects of Venture Capital Management Strategy

SCF was founded as part as VDI/VDE-Informationstechnik GmbH as an additional area of business activity. The latter had carried out the TOU pilot scheme, supporting NTBFs in the old Federal Republic of Germany, and started to launch a similar scheme in the new German Laender. Despite VDI/VDE-Informationstechnik's focus on information technology, the support program had to cover a broad range of traditional and emerging technologies. When the SCF started its business activities, it had a relatively broad range of investment opportunities, its primary aim being to support NTBFs.

The fund is relatively small, the capital raised amounts to DM 10 million, and the SCF uses capital sources provided by the BJTU program to refinance its investments. Without the public support programme, the BJTU pilot scheme, the SCF (as an additional operation of the VDI/VDE-Informationstechnik GmbH) would not have been founded in 1990. However, when the LBB came into the deal as the new owner of the SCF, the rather idealistic aim of supporting NTBFs gave way to that of making profits. In addition to this, the broad range of possible investments were reduced in favour of the more ambitious goal of earning money as it gained in experience. The SCF currently devotes most of its efforts to financing key technologies for industry.

The SCF now exists as an 'evergreen', independent of future state subsidies. State support influences investments strategies inasmuch as there are restrictions on investments with respect to a specific region (within Germany only) and depending on the extent of the investment (less than DM 1 million). As a small fund, the SCF is restricted to making investments with limited amounts of capital. Up to 1995 the fund had 19 investments with a total commitment of DM 6 million. All of the investments are in early-stage and first-rate company expansion. Since it started operations, two disinvestments have taken place, one involving a trade sale and the other a liquidation.

The Relationship between the Investment Company and the Portfolio Firms

MIOS

Not unlike other new technology based firms, MIOS started with a large capital requirement. In his position as chief executive of Logware, the head of the company maintained contact, in Berlin, with a corporate venture fund called the TIG (Technologie-Investitionsgesellschaft) with financial and industrial investors. The

TIG invested in Logware and favoured the idea of establishing a spin-out when it became clear that the two business activities within Logware demanded two distinct business areas. The SCF entered the deal when the TIG requested additional help to support the spin-out of a medium-sized software company. MIOS was one of the SCF's first investments. However, in terms of the previous experience of the head of the company, MIOS represented a typical start-up investment. Both the spin-out, MIOS and the SCF were new and the latter, in particular, only had a limited degree of experience in the venture capital market. Consequently, the previous experience and the syndication offered by the TIG helped to fix the deal. The talks with the SCF ran smoothly and resulted in a contract after only three meetings. The final agreement was based on a 49% share (24,5% for each venture capital company) as against 51% for the owner of MIOS. Due to the relatively small equity amount of MIOS, on the one hand, and the large capital requirement, on the other, the investment was split into an equity share and a silent share.

A financial bottleneck appeared in the fourth year of existence, when investments in equipment and machinery became necessary and, at the same time, a large customer from the automobile sector reduced his purchases. The SCF made a second round of financing possible without the TIG, which had closed its fund for new investments.

The management of MIOS by the SCF is characterized by active involvement and operational support in specific business areas. The SCF (and the TIG) provide consultancy in a number of company areas, for example: financing, marketing, organisation/staff recruitment, management.

The most important activity is financing, where MIOS depends on the ability of the SCF and the TIG to talk with other banks and extend bank loans, etc. Since the head of the company regards himself as a technical expert but lacks experience in management and business, he relies on external assistance in certain areas of business. On the recommendation of the SCF, he employed a young graduate marketing specialist as a business executive to concentrate on the company's core competence.

There are frequent contacts between all of the partners. They now hold intensive telephone conversations at least twice a week. Meetings are arranged every three months, and took place every two months during the first few years. However, apart from external assistance in business activities provided by the venture capital firms, the SCF and the TIG were unable to provide specific technical assistance. In the case of MIOS, technical know-how is a key area and continues to be important for expansion.

At present there is no clearly defined exit strategy. Both MIOS and the SCF have similar ideas about terminating the investment. However, their motivations are different. The head of the company expects it to grow considerably within the next year. Investments will be required which exceed the capacity of the current shareholders. Therefore a strong industrial partner that regards MIOS as a 'window on technology' would seem to be an adequate partner. From the point of view of the SCF, MIOS is one of the first investments and mature enough to harvest investments by making trade sales. MIOS would still seem to be too small to function as an IPO; only a secondary purchase by a large venture capital company could provide the opening needed by MIOS to go public or to enter the secondary market. A third round of financing is planned for the first few weeks of spring 1996.

Graphikon

In the early phase of its existence, Graphikon applied for public support for R&D. Due to the transformation process there were a number of service companies that were making support available to industrial companies in the new German Laender. Like other companies, Graphikon had to take care of its own financing, i.e. securing bank loans to fulfill the conditions for participation. The company was awarded several funds within the scope of the TOU program for the new German Laender, a programme which was carried out by VDI/VDE-Informationstechnik. At the same time, Graphikon's original bank displayed a growing reluctance to expand its credit line for further investments. After several years of participating in the TOU pilot scheme, Graphikon's founder and the SCF executive established their first contact at an investment forum. Whilst Graphikon was trying to solve its administrative problems with its home bank, another contact was established with a savings bank (the Sparkasse as part of the LBB).

Graphikon sought new financial ties that would respond more flexible and more comprehensively to the needs of company. The LBB with its subsidiaries, the Sparkasse and the SCF, was able to fulfill this need as it combined a traditional bank and an investment company providing business assistance. In 1994 a first round of five financing talks was held with the SCF. Because there had been personal contacts between the two relevant persons, the manager of the SCF and the head of the company before the talks took place, there seemed to be no major problems. However, the founder of Graphikon was hesitant to accept SCF investment in the form of a large equity share. He was interested in setting the equity share capital at 10%, whilst the SCF wanted to have a share anywhere between 10% and 15%. To assess the consequences of the SCF investment, the head of Graphikon called in an industrial manager to examine the contract and its impact on his company.

In October 1994 a contract was concluded between SCF and Graphikon. The equity share amounted to 10% and additional capital was provided in the form of a silent partnership. Graphikon also participated in all of the relevant public support programs for share capital investments (the TOU pilot scheme established for East Germany and the Innovationfund of the Berlin state). As a result, SCF venture capital has only played a minor role in the Graphikon's development and, at the same time, the SCF's influence on management decisions has been limited due to its relatively small share.

From the viewpoint of Graphikon, the support provided by the SCF facilitates the company's development. The problems and development of the company have been discussed at quarterly consultancy meetings. The SCF's management receives monthly business information from Graphikon. The consultancy activities which the SCF provides to Graphikon are comprehensive and range from establishing contacts to lawyers, tax consultants, etc. to brokering know-how and providing management consultancy (including the development of company strategies). Alongside the business assistance provided, Graphikon regards the involvement of the SCF's management an opportunity to learn through experience, i.e. to gain more knowledge about the specific field of technical business in which Graphikon operates. Furthermore, the SCF's ability to assist Graphikon in its future development is based on the investment company's external network. With regard to the relationship between the two companies, however, it seems uncertain whether more management and technical assistance would substantially improve the situation.

The head of Graphikon has reservations about a high-growth strategy. Instead of pursuing an explicit growth strategy, he is interested in stabilizing the company at the current employment level whilst simultaneously reducing the role played by public funds. Thus, from his point of view, further financing rounds for future expansion are irrelevant. Venture capital requires further development, which must sometimes be driven externally. This does not correspond with the ideas of the head of Graphikon for developing his company. Public grants for R&D, especially loans, are thus more adequate from his point of view, because external influences and control are thereby reduced. The company's turnover is still based on public grants running at around 40% of the total. For the head of Graphikon, it is still an open question as to how it will be possible to continue if SCF investments stop.

4.4   Specific Conclusions and Synopsis of the Empirical Evidence

The above observations are based on a sample of ten case studies, five from the US and five from Germany. As mentioned in the discussion of methodological preliminaries, it is, therefore, prudent to exercise extreme caution in generalizing the research findings established to date and to avoid drawing more than a set of tentative conclusions. It seems rather difficult to present a clear distinct picture of each VC industry. As Roberts has pointed out, this industry is highly dynamic and stereotypes should be avoided.[114] Alternative interpretations may become more plausible in the light of further survey data, which should necessarily include a broader range of both successful and unsuccessful investments in NTBFs in each country.

Despite these caveats, some of the findings in the case studies do seem worth mentioning. The major distinction between the literature screening and the case studies is the level of investigation. Whereas the literature review discussed the key differences between the US and German VC industry at the aggregate level, especially the framework conditions, the empirical analysis allows us to present some findings on the micro level.

A first finding relates to the ownership structure within a venture backed NTBF and the kind of management involvement during the course of investments by a VCC. Proceeding from the differentiations made by Macmilllan, Kulow and Khoylian[115] it is possible to identify three distinct levels of involvement adopted by venture capitalists:
-     laissez faire involvement, in which the venture capitalist displays limited involvement

---

[114] Cf. Roberts 1991b, p.12

[115] Cf. MacMillan et al. 1988

- moderate involvement, in which the venture capitalist displays moderate involvement
- close-tracker involvement, in which the venture capitalist displays more involvement than the entrepreneur in the lion's share of his business activities

In all of the above cases - with the exception of MBG Hessen which allocates silent partnerships - the US and German venture capitalists provided hands-on investment rather than laissez faire involvement. However, in US portfolio companies, the investment firm acts more as a substitutive management within the firm and therefore exhibits 'close-tracker' qualities. In German portfolio companies, by way of contrast, venture capitalists often act as an external complement to the internal management of the company. Hence, the activities of the German portfolio companies would appear to be better characterized as moderate involvement.

The case of AccessLine/Apex in the US, for instance, showed that managerial incentives are crucial for the investment decisions of US venture capitalists. The example of ImmuLogic/Greylock demonstrate that beyond the goal of assuring NTBFs progress, the maximization of control over the firm's equity and the assimilation of a wide range of influences in the NTBF is also important. The example of Documentum/XTV showed a specific fund strategy which included hiring outside management and taking active roles in the supervision and manage-ment of portfolio firms as board members and informal advisors. The case of ISC, where the head of ARCH Venture Partners became company's chairman of the board of directors, underlines the very active involvement of US venture capitalists. However, the example of RhoMed/Aberlyn demonstrate, that in some cases US venture capitalists are satisfied with a more passive role of limited engagement.
It cannot be said that German investments take place without external managerial involvement when investments are made. As in the US, the role of informal advisors is always translated into action, with the exception of MBG Hessen, which became involved in BioTec on a laissez faire basis. The cases of MIOS/SCF and

Graphikon/SCF, however, appear to be characterized less by vigorous action than by a good management partnership similar to the practice of US venture capitalists. In the case of Tomtec/Atlas, however, with a syndication of seven VC companies, a coherent management strategy is hard to find, despite the quarterley „board meetings" that come close to US board meetings.[116]

A second finding concerns the management abilities and business expectations of entrepreneurs. This has to be seen within the context of the managerial behaviour of venture capitalists. As Gorman and Sahlman have pointed out, in the US both entrepreneurs and venture capitalists alike consider accepting venture capital as equivalent to entering a partnership.[117] In Germany the founders of an NTBF generally wish to remain sole proprietors even if this limits the growth prospects of a firm. This basic position has a high degree of acceptance with respect to venture capital, especially in the case of open shares.[118]

Cultural and personal determinants are always mentioned in connection with Germany, whereas US entrepreneurial history reveals the powerful influence of the pioneering spirit and the striving for success. The striving for independence and the slight acceptance, by founders of firms and of partners with equal rights may, however, influence the conception and evaluation of venture capital as a means of financing NTBFs from the view of an enterpreneur. The case of Graphikon and the SCF provides a good example of the motivational structure of German entrepren-

---

[116] One possible explanation could be found in the Board of Directors, an institution within US companies that does not have an exact equivalent in German companies. Gorman and Sahlman (1989, p.241) indicate that venture capitalists, acting through the Board of Directors, typically gain power to hire and fire management. As mentioned in chapter 3.3 even as minority investors venture capitalists will usually occupy at least one board seat in a portfolio firm. In discussions with German VC managers, this topic was mentioned sometimes; owing to the completely different situation in Germany. However, the degree to which having a seat on the Board of Directors might be important for a venture capitalist would seem to be highly speculative matter.

[117] Gorman/Sahlman 1989, p.241

[118] Cf. Ludsteck 1994

eurs, which resulted in interests diverging from those of the venture capitalists, and had consequences for participation agreements. In the case of BioTec and the MBG Hessen these types of problems were not supposed to arise due to the broader business orientation of the MBG Hessen. The differences between the values and the abilities of German firms and managers are clearly narrow when seen on an international level.[119] This would still appear to apply to founders of firms. In the US, a high income and the striving for high achievements are the main motives behind establishing firms. In the Federal Republic, however, this move is dominated by the desire to increase personal decisions, the freedom of action as well as independence and the attainment of personal goals. These different attitudes may lead to various forms of conflicts and, at the very least, result in divergent starting points if one compares US and German venture capitalists in the initial stage of negotiating contracts.

A third finding reveals that levels of business experience are sometimes different when one compares German and US investment managers, e.g. the knowledge about the idiosyncrasies of the technology and the sector. In the case of Immulogic, for example, the venture capitalist of Greylock had acquired extensive experience in the relevant biotechnology sector. In the case of ISC, the venture capitalist who headed the board of directors came as president, and CEO from ARCH development corporation, the general partner of ARCH venture partners. He was the driving force behind the development of this university related technology transfer organization and established a very impressive track record by founding a dozen firms, generating $2 million worth of royalties for the University of Chicago and Argonne Laboratories. The most obvious German example is provided by the case of EOS, where the relevant venture manager at Technologieholding had previously run his own company in the industry where the proposed investment was to be made.

---

[119] See for example, Getas 1989; Grimm 1985; Mohler 1989.

There are two possible explanations for the greater experience of US venture capitalists. Firstly, the degree of specialization with regard to their industrial investments is based on past experience, generally from the same industries and from current contacts with suppliers, customers and engineers related to this industry. Secondly, despite an growing trend for the venture capital industry to become increasingly averse to taking risks, high-tech industries such as biotechnology and computer software, for example, are of greater importance for investments than traditional industries. In contrast, German VC managers frequently do not specialize in certain high-tech sectors or branches. They prefer a broad distribution of industries in their portfolios to reduce the possible risk of failures. Thus the strategy adopted holds the advantage of not relying on developments within one specific sector. On the other hand, and this is of greater importance, the less the degree of specialization, the more difficult it is to realize the benefits acquired through learning, e.g. the knowledge and experience gained about a specific industry.[120]

The following synopses contain the main features of the case studies described above.

---

[120] Cf. Wupperfeld 1993, p.71

Table 17	Synopsis of US Case Studies				
Company name	AccessLine	Immulogic	Docu-mentum	Illinois Super-conductor	RhoMed
Venture Capital fund	APEX- Invest-ment	Greylock	Xerox Tech-nology Venture	ARCH Venture Partners	Aberlyn Capital Management
Formation of the company	1989	1987	1990	1989	1986
Background of entrepre-neur	engineer with several years of business experience in the area of mobile electronics and telephone communica-tions	Professor at MIT since 1977	engin-eering execu-tives from a database manufac-turer	scientists from Argonne Laboratories and University from Chicago	scientist with several years of business experience
The techno-logy	smart number applications/ telecommuni-cations	immunology/ biotechno-logy	object-oriented document manage-ment system/ software	high-temperature supercon-ductors/elec -tronics	nuclear medicine/ biotechno-logy
Phase of investment	start-up	later stage	seed	seed	later stage
Means of exit	not decided yet	IPO	IPO	IPO	not decided yet
Key issues to emerge	entrepreneur founded his first company in 1976; experienced management team; VC company demanded influ-ence on mana-gerial decisions early on	VC manager with exten-sive experi-ence in biotechno-logy	spin-out assisted by VC company; parent company provided certifi-cation and cus-tomer network	spin-off by VC company; head of VC company chairman of the Board of Directors; awarded funds from SBIR program	awarded sev-eral public grants; new VC strategy through Venture leas-ing; limited engagement; VC company kept own technical experts

Table 18	Synopsis of German Case Studies				
Company name	**EOS**	**Tomtec**	**BioTec**	**MIOS**	**Graphikon**
Venture Capital fund	Technolo-gie-holding	Atlas Venture	MBG Hessen	Seed Capital Fund	Seed Capital Fund
Formation of the company	1989	1990	1986	1990	1990
Background of entrepreneur	physicist with several years of business experience	technician with several years of business experience	chemist	software engineer and executive in software company	scientist and head of department at Academy of Science (GDR)
The technology	laser based prototyping/ laser technology	3-D-echo computer tomography / medical technology	in-situ detoxification /environment technology	wireless and infra-red data communica-tion/ software	image processing/ software
Phase of investment	start-up	start-up	start-up	start-up	start-up
Means of exit	trade sale	not decided yet	not decided yet (buy back possible)	not decided yet (trade sale possible)	not decided yet
Key issues to emerge	awarded several public grants; network to large corporate customers; negotiations with VC company led to liquidity shortages	awarded several public grants; net-work to large corporate customers; acquisition of further capital for expansion not suc-cessful	awarded public grants from TOU program; silent part-nership	spin-out by medium sized software company; awarded public grants; expansion of company comes up to the limits of the seed company	spin-off from ADW; support from public grants; intensive negotiations with seed company about equity share and managerial influence; rather future stabilization than future expansion

5   Summary and Overall Conclusions

This US-German study has compared the role of venture capital in these two countries as a means of financing new technology based firms. Well-known examples of today's large businesses in the US, for example, Apple, Advanced Micro Devices, Digital Equipment and Intel are often selected to demonstrate that, in some cases, venture capital has had a significant impact on NTBFs. In Germany, discussions about venture capital exerting this kind of impact have been audible since the mid-eighties. However, while the differences between the US and German VC markets seemed to have narrowed over time, one can still point out several distinctions between venture capital industries in the US and Germany.

Despite the obvious differences, for instance
- the size and rate of growth of the VC industries
- the legal structure of venture capital funds, and
- the market structure of both VC industries

the comparison identified several other differences that seem to important enough to deserve mention.

The results of the literature screening clearly indicate the existence of predominantly positive development factors in the US.

The first development factor is the relatively long tradition of venture capital in the US. VC started in 1946 with the foundation of ARD. This has resulted in a large number of highly professional VC managers. The second development factor consists of a combination of direct and indirect policy initiatives. These initiatives include policy schemes such as the implementation of Small Business Investment

Centers (SBICs) in 1958, which paved the way for broader involvement by venture capital institutions. Furthermore, it has to be pointed out that indirect policy initiatives, for example, acts and regulations - such as the ERISA "Prudent Man" Rule, the Small Business Investment Act, and the ERISA "Safe Harbour" Regulation - have created a stable framework for investors and investees and provided a growing supply of money for VC activities. A third factor can be categorized as exogenous. The conditions for successful integration might appear to be superior to those in Germany. Developments such as technological break-throughs in microelectronics and biotechnology and the acquisition of NTBFs by large Japanese and US companies are regarded as having helped the growth process of venture capital in the United States.

The German framework conditions that have been taken into account here mainly show development hurdles which allegedly hinder the development of the venture capital market. It seems important to draw attention to three such hurdles. Firstly, the German universal banking system and the dominant role of the credit institutes in the financing of young and medium-sized firms. Secondly, the lack of opportunities for disinvesting within the framework of going public. IPOs, in particular, are of minor relevance in comparison with the opportunities provided by both the NASDAQ and the relevant entry and valuation regulations. The youngest and most willing to take risks are generally those most effected by this situation. VC investments are insufficient to realize high dividends for NTBFs. As a consequence, there does not exist an adequate capital market segment for NTBFs in Germany to prevent the exit of investment companies there. The third hindrance concerns the lack of preferential tax treatment for share capital. In connection with the second development hurdle, it would seem rational for the investor to find later-stage investments more attractive. Early-stage investments, as discussed in the introductory outline, pose higher risks due to the uncertainty surrounding firms in their development stage. Thus, a larger amount of private and corporate capital is allocated to those types of investments which offer higher rates of return or involve lower risks on involvement such as, for example, later-stage investments or completely different capital investments.

With regard to the history of the German VC industry, it has already been stated that a long-term venture capital tradition does not exist in the Federal Republic of Germany as it does in the US. It was, therefore, only in the eighties that a market volume worth mentioning came into existence. Even in this case, however, most companies making share investments concentrated their activities on financing the growth processes of established firms, e.g. MBO. Due to this absence of a VC tradition and the short period of existence of the venture capital market, there is still a shortage of qualified professionals in Germany today. Up to now, only a few investment capital companies have all-round experience with NTBFs. The efficient networks and suitable instruments for the evaluation, management and control of NTBFs required in order to achieve higher dividends have consequently only been developed recently by a few specialized companies. In contrast to the US, there is no "business angel community" in Germany.[121]

However, as in the US, one development factor has provided some important impulses for the development of the VC market: if one considers the development of the German venture capital market, one can see that the situation for VC and NTBFs has clearly become more favourable thanks to the encouragement given by public policy and, in particular, the support provided by the BJTU pilot scheme as well as the measures it has taken so far, for example, the BTU pilot scheme that provides share capital investments for NTBFs and the TOU pilot scheme for NTBFs. There is a widespread agreement that the German venture capital market for NTBFs would be far less important without such a massive support.

---

[121] The expert discussion at the workshop emphasized this aspect but could not provide any substantial information.

Apart from taxation, legal and economic factors, the development of venture capital markets is also influenced by social and cultural factors. These factors are quite difficult to asses with regard to their impact on VC markets. Furthermore, very little research has been done in this field, especially in the US, and the investigation approach adopted has not been able to provide answers in this multi-disciplinary area of investigation into venture capital. Nevertheless, when the role of entrepreneurship in the US and environmental conditions are discussed, socio-economic factors are often mentioned. Hence, any comprehensive survey of a national framework should at least mention these factors.

One socio-economic factor that is often emphasized is the mentality factor, i.e. the national interpretation of value as well as of firms and managers. It has been mentioned that the US creates better conditions for venture capital. Empirical analyses show that the mentality of people in the US is fundamentally different from that of the Europeans, and from that of the Germans in particular.[122] These analyses confirm the picture of a safety- and hierarchy-oriented German who also attaches great importance to consumption and leisure-time. In the US, however, performance and career ambitions seem to be significantly more developed, which is not surprising considering US history, which is influenced by the pioneering spirit and the striving for success. Added to this, the frequently discussed unwillingness of the Germans to take risks, and the close social network associated with employee status hinder the transfer of management talents in small firms.[123]

The case studies undertaken were not able to provide evidence on the influence of mentality on VC decisions with respect to portfolio firms and investment companies.[124] Personal determinants such as the striving for independence and the

---

[122] See for instance, Fetzer 1990; Grimm 1985; Hierl 1984; Ludsteck 1993

[123] Cf. Fetzer 1990; Grimm 1985; Hierl 1984; Ludsteck 1993

[124] Although the case studies revealed no direct evidence of this phenomenon, and became apparent in the expert discussion at the workshop that this aspect was repeatedly referred to as a development problem as far as the German VC market was concerned.

minor acceptance by firm founders of partners with equal rights have been mentioned by some of the interviewed German entrepreneurs, and may have determined their final decisions about investment contracts. On the other hand, however, the involvement of German venture capitalists does not always seem to be accompanied by vigorous activities. Given the different degrees of involvement of US and German venture capitalists as managers within the scope of their portfolios, the latter do not appear to be such `close-trackers´ as their US counterparts.

From the case studies and the literature reviewed, additional managerial differences between the US and German venture capitalists became apparent. Firstly for US managers, the degree of specialization with regard to their industrial investments is based on past experience, mostly within the same industries and through current contacts with suppliers, customers and engineers related to these industries. Secondly, despite an increasing tendency for the US venture capital industry to become more averse to risks, high-tech industries such as biotechnology and computer software, for example, still remain more important for investors than the traditional industries. As mentioned above, German VC managers frequently do not specialize in certain high-tech sectors or branches. They prefer a broad distribution of industries in their portfolios in order to reduce possible risks of failure. Thus the strategy adopted contains the advantage of not being dependent on developments within a specific sector. On the other hand, and this is of greater importance, the less the degree of specialization, the more difficult it is to realize the benefits gained through learning, e.g. knowledge and experience about a specific industry.

Finally, evidence from previous and the current investigations has shown that US venture managers seem to differ from their German counterparts in several ways. With all these statements however, one must bear in mind the particular historical developments and framework conditions for the venture capital market in the US and Germany. An analysis or judgement of venture capital industries would not appear to make sense unless it specifically refers to these aspects. The differences in the performance and involvement of VC managers in NTBFs, as identified by the case studies, cannot, however, be judged in terms of differences in importance. The empirical basis of this investigation was too narrow to draw much more than tentative conclusions. The task of this study was not to provide recommendations on how this situation can be changed. It is, however, doubtful whether the US venture capital market is a relevant measure of size and success. At most, the German venture capital industry ought to carefully investigate what it can learn from the US experience.

6   References

Albach, H., Husdiek, D., Kokalj, L., 1986, "Finanzierung mit Risikokapital", Poeschel, Stuttgart.

Amit, R., Glosten, L., Muller, E., 1990, "Does Venture Capital Foster the Most Promising Entrepreneurial Firms?", in: California Management Review, pp. 102-111.

Bachelier, R., 1993, "Die neuen Fördermöglichkeiten des BMFT zur Finanzierung des Produktionsaufbaus und der Markterschließung" (Phase II des Modellversuchs TOU-NBL), in: Bräunling, G.; Pleschak, F.; Sabisch, H. (Red.): Finanzierung des Produktionsaufbaus und der Markterschließung geförderter junger Technologieunternehmen in den neuen Bundesländern, Konferenz und Workshop am 16. und 17. März 1993 in Leipzig, Tagungsbericht, Fraunhofer-Institut für Systemtechnik und Innovationsforschung (ISI), Karlsruhe, S. 4-5.

Bannock, G. and Partners Ltd, 1994, "European Second-Tier Markets for NTBFs", Study carried out for the European Commission DGXIII-D4: Sprint/EIMI, Brussels-Luxembourg.

Barry, C., Muscarella, C., Peavy, J., Vetsuypens M., 1990, "The Role of Venture Capital in the Creation of Public Companies: Evidence from the Going Public Process", in: Journal of Financial Economics, 27, pp. 447-472.

Barry, C., 1994, "New Directions in Research on Venture Capital Finance", Journal of Financial Management, Volume 23, No.3, Autumn, pp. 3-15.

Bhide, A., 1992, "Bootstrap Finance: The Art of Start-Ups", Harvard Business Review, (Nov. - Dec. 1992).

Birch, D., 1990, "Sources of Job Growth and Some Implications", in: Jobs, Earnings, and Employment Growth Policies in the United States, ed. J. Kasarda, Kluwer Academic Publishers, Norwell, MA, pp. 71-76.

Block, Z., MacMillan, I., 1993, "Corporate Venturing", Harvard University School Press, Boston.

Bräunling, G., Gerybadze, A., Mayer, M., 1989, "Ziele, Instrumente und Entwicklungsmöglichkeiten des Modellversuchs "Beteiligungskapital für junge Technologieunternehmen" (BJTU) ", Arbeitspapier des Fraunhofer-Instituts für Systemtechnik und Innovationsforschung (ISI), Karlsruhe.

Bräunling, G., Pleschak, F., Sabisch, H. (Red.), 1993, "Finanzierung des Produktionsaufbaus und der Markterschließung geförderter junger Technologieunternehmen in den neuen Bundesländern", Konferenz und Workshop am 16. und 17. März 1993 in Leipzig, ISI, Karlsruhe.

Bruno, A.V., Tyebjee, T.T., 1982, "The Environment for Entrepreneurship", in: Kent, C.A. et al. (Eds.): Encyclopedia Of Entrepreneurship, Englewood Cliffs; pp. 288-307.

Büschgen, H.E., 1985, "Banken und Venture-Capital-Finanzierung", in: Die Bank Nr. 6, S. 284-292.

Butchart, R.L., 1987, "A New UK Definition of the High Technology Industries", in: Economic Trends, No. 400, pp. 82-88.

Bygrave, W. D. Timmons, J. A., 1992, "Venture Capital at the Crossroads", Harvard Business School Press, Boston.

BVK (ed., various issues), Bundesverband Deutscher Kapitalbeteiligungsgesellschaften, Jahrbuch, Berlin.

Churchill, N.C., Lewis, V.L., 1983, "The Five Stages of Small Business Growth", in: Harvard Business Review, Vol. 33, No. 3, pp. 30-50.

Clark, A. S., 1994, "Technology Policy and Venture Capital", Financing Entrepreneurs, American Enterprise Institute for Public Policy Research, Washington D.C., pp. 45-50.

Cooper, A.C., Bruno, A.V., 1977, "Success Among High-Technology Firms", in: Business Horizons, Vol. 20, April, pp. 16-22.

Cooper, A.C., Dunkelberg, W.C., 1981, "A New Look at Business Entry: Experiences of 1805 Entrepreneurs", in: Vesper, K.H. (Ed.): Frontiers of Entrepreneurship Research, Wellesley, pp. 1-19.

Cooper, A.C., 1981, "Strategic Management, New Ventures and Small Business" in: Long Range Planning, Vol. 14, No. 5, pp. 39-45.

Cooper, A.C. 1986, "Entrepreneurship And High Technology", in: Sexton, D.L.; Smilor, R.W. (Eds.): The Art and Science of Entrepreneurship, Cambridge, pp. 153-168.

Dean, B., Giglierano, J.J., 1989, "Patterns in Multi-Stage Financing in Silicon Valley", in: Vancouver College on Innovation Management and Entrepreneurship: Proceedings of Vancouver Conference, May 1989.

Dubini, P., 1988, "The Influence of Motivations and Environment on Business Start-Ups: Some Hints for Public Policies", in: Journal of Business Venturing, Vol. 4, pp. 11-26.

Dunkelberg, W.C., Cooper, A.C., 1983, "Financing the Start of a Small Enterprise", in: Hornaday,J.A.; Timmons, J.A.; Vesper, K.H. (Eds.): Frontiers of Entrepreneurship Research 1983, Wellesley, pp. 369-381.

Eisenhardt, K.M., Forbes, N., 1984, "Technical Entrepreneurship: An International Perspective", in: Columbia Journal of World Business, Vol. 19, No. 4, pp. 31-38.

Europe's venture capital Association (EVCA), 1992, Venture Capital in Europe, 1991 EVCA Yearbook, London.

Europe's venture capital Association (EVCA), 1993, Venture Capital in Europe, 1992 EVCA Yearbook, London.

Fazzari, G.R. et al., 1988, "Financial Constraints and Corporate Investment", Brookings Papers on Economic Activity 1, pp.141-195.

Fendel, A., 1987, "Investmententscheidungsprozesse in Venture Capital - Unternehmungen: Darstellung und Möglichkeiten der intrumentellen Unterstützung", Köln.

Fenn, G. Liang, N., Prowse, S., 1995, "The Economics of the US Private Equity Market." Draft Report, Board of Governors, Federal Reserve of the United States.

Fetzer, R., 1990, "Analyse internationaler Unterschiede im Volumen und in der Struktur von Venture-Capital-Aufkommen und -Anlage, Analysen zur Strategie ausgewählter Akteure im Netzwerk der jungen Technologieunternehmen", Berlin.

Fischer, L., 1987, "Problemfelder und Perspektiven der Finanzierung durch venture capital in der Bundesrepublik Deutschland", in: Die Betriebswirtschaft, Jg. 47, Nr. 1, S. 8-32.

Florida, R., 1994, "Keep Government out of Venture Capital", Financing Entrepreneurs, American Enterprise Institute for Public Policy Research, Washington D.C., pp. 51-57.

Florida, R., Kenney, M., 1988, "Venture Capital - Financed Innovation and Technological Change in the US", in: Research Policy, Vol. 17, No. 3, pp. 119-137.

Florida, R., Smith, D.F., 1990, "Venture Capital, Innovation and Economic Development", in: Economic Development Quarterly, Vol. 4, No. 4, pp. 345-360.

Florida, R., Smith, D.F., 1993, "Venture Capital and Industrial Competitiveness", Economic Development Administration.

Freear, J., Wetzel, W.E., 1990, "Who Bankrolls High-Tech Entrepreneurs?", Journal of Business Venturing. Volume 5, No. 2. March, pp. 77-89.

Freear, J., Wetzel, W.E., 1992, "The Informal Venture Capital Market in the 1990s", in: Sexton, D.L.; Kasarda, J.D. (Eds.): The State of the art of Entrepreneurship, Boston, pp. 462-486.

Frommann, H., 1992, "Venture Capital in Deutschland - Rückblick auf ein Vierteljahrhundert", in: Bundesverband deutscher Kapitalbeteiligungs- gesellschaften (Hrsg.): Geschäftsbericht 1992, S. 101-106.

Frommann, H., 1993, "Entwicklungstrends am deutschen Beteiligungsmarkt", in: Bundesverband deutscher Kapitalbeteiligungsgesellschaften (Hrsg.): Geschäftsbericht 1993, S. 11-31.

Galbraith, J.R., 1982, "The Growth of New Business Ventures", in: Ansoff, H. et al. (Eds.): Understanding and Managing Strategic Change, Amsterdam, New York, Oxford, pp. 63-81.

Gerybadze, A., 1988, "The Organizational Life Cycle of NTBFs", in: Anglo-German Foundation (Ed.): New Technology-Based Firms in Britain and Germany, London, pp. 55-73.

Gerybadze, A., 1991, "Marktwirtschaft und innovative Unternehmensgründungen". Erfahrungen aus dem Modellversuch "Förderung technologieorientierter Unternehmensgründungen" (TOU), in: Oberender, P; Streit, M.E. (Hrsg.): Marktwirtschaft und Innovation, Baden-Baden, S. 123-157.

Getas, 1989, "Die Deutschen als Europäer - Teilergebnisse aus ACE Anticipating Change of Europe und Getas-Report", Hamburg.

Giersch, H., 1982, "Ausbruch aus der Stagnation - Chancen für neue Arbeitsplätze", in: Risiken und Chancen der künftigen Wirtschafts-entwicklung, Kieler Diskussionsbeiträge 84, Kiel, S.3-8.

Gillner, G., 1984, "Eigenkapitalfinanzierung technologischer Neuentwicklungen, eine Analyse von Venture Capital - Fonds in der Bundesrepublik Deutschland", Arbeitspapier Nr. 7 des Instituts für Wirtschaftswissenschaften der TU Braunschweig, Braunschweig.

Gompers, P., 1994a, "Grandstanding in the Venture Capital Industry," University of Chicago, mimeo.

Gompers, P., 1994b, "Optimal Investment, Monitoring, and the Staging of Venture Capital," University of Chicago, mimeo.

Gompers, P., Lerner, J., 1994, "The Structure of Compensation in the US Venture Capital Partnership", Harvard Business School, Working Paper 95-009.

Gorman, M., Sahlman, W.A., 1989, "What do Venture Capitalists do?", in: Journal of Business Venturing 4, pp. 231-247.

Goslin, L.N., Barge, B., 1986, "Entrepreneurial Qualities Considered in Venture Capital Support", in: Ronstadt, R. et al. (Eds.): Frontiers of Entrepreneurship Research 1986, Wellesley, pp. 366-379.

Greiner, L.E., 1972, "Evolution and Revolution as Organizations Grow", in: Harvard Business Review, July-August.

Grimm, E., 1985, "Wertewandel-Konsumwandel", in: Planung und Analyse, Jg. 12, Nr. 9, S. 392-396.

Grisebach, R., 1989, "Innovationsfinanzierung durch Venture Capital - Eine juristische und ökonomische Analyse", München.

Gupta, A.K., Sapienza, H.J., 1988, "The Pursuit of Diversity by Venture Capital Firms: Antecedents and Implications", in: Kirchhoff, B.A. et al. (Eds.): Frontiers of Entrepreneurship Research, Wellesley, pp. 290-302.

Gupta, U., 1986, "Hands-On Venture Capital", in: Venture, Jan., pp. 45-50.

Gupta, U., 1990, "Venture Capital Dims for Start-ups, but not to Worry," Wall Street Journal, January 24.

Gupta, Y.P., Chin, D.C.W., 1993, "Strategy Making and Environment: An Organizantional Life Cycle Perspective", in: Technovation, Vol. 13, No. 1, pp. 27-44.

Halloran, M., L. Benton, Lovejoy, J., 1992, "Venture Capital and Public Offering Negotiation", Harcourt Brace Jovanovich, New York.

Harnischfeger, M., Kulicke, M., Wupperfeld, U., 1992, "Zum Stand des Modellversuchs "Beteiligungskapital für junge Technologieunternehmen" (BJTU) " - Zwischenbericht zum 31.12.1991, Arbeitspapier des Fraunhofer-Instituts für Systemtechnik und Innovationsforschung (ISI), Karlsruhe.

Harrison, R.T., Mason, 1991, "Informal Investment Networks: a Case Study from the United Kingdom", in: Entrepreneurship and Regional Development, Vol. 3, pp. 269-279.

Hart, S.L., Denison, D.R., 1987, "Creating New Technology-Based Organizations - A System Dynamics Model", in: Policy Studies Review, Vol. 6, pp 512-528.

Hausberger, H., 1984, "Wiederbelebung der Aktie", in: Wirtschaftsdienst, Nr. 7, S. 335-340.

Hierl, W., 1984, "Venture Capital - auf deutsche Verhältnisse übertragbar? ", in: Kreditpraxis, Nr. 3, S. 35-46.

Horvath, P., Winderlich, H.G., Zahn, E., 1984, "Unternehmensgründungen in Bereichen der Spitzentechnologie", in: Albach, H.; Held, T. (Hrsg.): Betriebswirtschaftslehre mittelständischer Unternehmen, Stuttgart, S. 133-147.

Huemer, J., 1992, "Public Venture Capital," Venture Capital Journal, February, 39.

Ingram, D.H.A., Miles, D.K., 1984, "Unternehmensfinanzierung in Großbritannien und in der Bundesrepublik Deutschland", in: Monatsberichte der Deutschen Bundesbank, Nr. 11, S. 35-46.

Jansson, S., 1984, "The Leap of Faith into Venture Capital", Institutional Investor, September, pp. 117-121.

Jensen, M., 1989, "Eclipse of the Public Corporation", Harvard Business Review, 5, S. 61-74.

Kaplan, S., 1991, "The Staying Power of Leveraged Buy-outs," Journal of Financial Economics, 29, pp. 287-314.

Kasarda, J., Sexton, D., 1992, "The State of the Art of Entrepreneurship", PWS-Kent Publishing Company, Boston.

Kazanjian, R.K., 1984, "Operationalizing Stage of Growth: An Empirical Assessment of Dominant Problems", in: Hornaday, J.A. et al. (Eds.): Frontiers of Entrepreneurship Research 1984, Wellesley, pp. 144-158.

Kotkin, J., 1984, "Why Smart Companies are Saying NO to Venture Capital," INC, August, pp. 65-75.

Kazanjian, R.K., Drazin, R., 1989, "An Empirical Test of a Stage of Growth Progression Model", in: Management Science, Vol. 35, No. 12, pp. 1489-1503.

Klemm, H.A., 1988, "Die Finanzierung und Betreuung von Innovationsvorhaben durch Venture Capital Gesellschaften. Möglichkeiten und Grenzen der Übertragung des amerikanischen Venture Capital Konzepts auf die Bundesrepublik Deutschland. Frankfurt am Main.

Kulicke, M., 1987, "Technologieorientierte Unternehmen in der Bundesrepublik Deutschland - eine empirische Untersuchung der Strukturbildungs- und Wachstumsphase von Neugründungen", Frankfurt am Main, Bern, New York.

Kulicke, M., Gerybadze, A., 1990, "Entwicklungsmuster technologieorientierter Unternehmensgründungen", Karlsruhe.

Kulicke, M., u.a., 1993, "Chancen und Risiken junger Technologieunternehmen - Ergebnisse des Modellversuchs "Förderung technologieorientierter Unternehmensgründungen" (TOU) ", Heidelberg.

Kulicke, M., 1995, "Hintergrundinformationen zur Pressekonferenz von Bundesminister Dr. Rüttgers anläßlich der Vorstellung des neuen Förderprogramms "Beteiligungskapital für kleine Technologieunternehmen" (BTU) ", Arbeitspapier des ISI, Karlsruhe.

Lampe, D., 1992, "Route 128: Lessons from Boston's High-Tech Community", Basic Books, New York.

Lerner, J., 1994a, "The Syndication of Venture Capital Investments", Financial Management, 23 (Autumn).

Lerner, J., 1994b, "Venture Capitalists and the Decision to Go Public", Journal of Financial Economics, 35, pp. 293-316.

Lerner, J., 1995, "Venture Capitalists and the Oversight of Private Firms", Journal of Finance, 50, forthcoming.

Liles, P., 1974, "Sustaining the Venture Capital Firm, Management Analysis Center Research Study, Homewood (Ill.).

Ludsteck, W., 1993, "Lohnt sich das Wagnis der Wagnisfinanzierung?", in: Süddeutsche Zeitung, 14.9.1993, S. 21.

Ludsteck, W., 1994, "Risikoscheu macht Risikokapital zu schaffen", in: Süddeutsche Zeitung, 21.11.1994, S. 23

MacMillan, I., Kulow, D.M., Khoylian, R., 1988, "Venture Capitalists' Involvement in their Investments: Extent and Performance", in: Kirchhoff, B.A. et al. (Eds.): Frontiers of Entrepreneurship Research, Wellesley, pp. 303-323.

MacMillan, I.C., Siegel, R., Narasimha, P.N.S., 1985, "Criteria Used by Venture Capitalists to Evaluate New Venture Proposals", in: Journal of Business Venturing, Vol. 1, pp. 119-128.

Mason, C., Harrison, R., 1992, "The Financing of Technology-Based New Firms in the UK", In: ISI, Warwick Business School (Eds.): Proceedings of the Anglo-German Seed-Capital Workshop, Karlsruhe, Warwick, pp. 58-86.

Mayer, M. et al., 1989, "Modellversuch "Förderung technologieorientierter Unternehmensgründungen (TOU)", Zwischenbilanz zum 31.12.1988, Karlsruhe.

Mayer, M., Müller, R., 1991, "Die deutsche Wagnisfinanzierungsgesellschaft mbH (WFG) - Erfahrungen und Ergebnisse eines Modellvorhabens", Arbeitspapier des ISI, Karlsruhe.

McMurty, B.J., 1986, "Tax Policy Influence on Venture Capital", in: Landau, R.; Jorgenson (Eds): Technology and Economic Policy, Cambridge, pp. 137-151.

Mitchell, J.C., 1983, "Case and situational analysis, Sociological Review", 31, pp. 187-211.

Mohler, P.P., 1989, "Der Deutschen Stolz. In: Informationsdienst Soziale Indikatoren, Nr. 2, S.1-4.

Müller-Kästner, B., 1989, "Das Mittelstandsprogramm soll größenbedingte Nachteile bei der langfristigen Fremdfinanzierung ausgleichen", in: Handelsblatt, 8. April, S. 34.

Murray, G., 1992, "Exit Problems for Seed and Early Stage Venture Capitalists: the Second "Equity Gap", The European Foundation for Entrepreneurship Research, 5th Annual European Forum, London Dec.

Murray, G., Francis, D., 1992, "The European Seed Capital Fund Scheme: Review of the First Three Years", Warwick Business School, Coventry.

Murray, G., 1995, "A Synthesis of Six Exploratory", European Case Studies of Successfully-Exited, Venture Capital Financed, New Technology Based Firms, mimeo.

Nevermann, H., Falk, D., 1986, "Venture Capital", Baden-Baden.

Pettigrew, A., 1973, "The Politics of Organisational Decision Making", Tavistock, London.

Pfirrmann, O., 1994, "The Promotion of New Technology Based Firms in the New German Laender: A New Innovation Support Approach", in: Technology Transfer Practice in Europe, (TII and SPRINT, eds.), Vol.I, Luxembourg.

Pichotta, A., 1990, "Die Prüfung der Beteiligungswürdigkeit von innovativen Unterneh-mungen durch Venture Capital-Gesellschaften", Bergisch Gladbach, Köln.

Picot, A., Laub, U.D., Schneider, D., 1989, "Innovative Unternehmens-gründungen. Eine ökonomisch-empirische Analyse", Berlin, Heidelberg, New York, London, Paris, Tokyo.

Pollack, A., 1989, "Venture Capital Loses its Vigor", New York Times, October 8.

Poterba, J., 1989, "Venture Capital and Capital Gains Taxation", Tax Policy and the Economy, 3, pp. 47-67.

Pratten, C., 1993, "The Stock Market", Cambridge.

Premus, R., 1984, "Venture Capital and Innovation", Report to the Joint Economic Committee Congress of the United States, Washington: Government Printing Office..

Price Waterhouse, 1992, "Software Industry Business Practices Survey", Massachusetts Computer Software Council, Inc. and Price Waterhouse.

Quillmann, W., 1987, "Venture Capital in den US und Deutschland", in: Die Bank, Nr. 12, S.669-673.

Retkwa, R., 1990, "Venture Industry Now in Transition Period", Pension World, July, pp.24-26.

Riesenhuber, H., 1984, "Wagnisfinanzierung und technologieorientierte Unternehmensgründung, St. Augustin.

Roberts, E.B., 1970, "How to Succeed in a New Technology Enterprise", in: Technology Review, December, pp. 23-27.

Roberts, E.B., 1991a, "Entrepreneurs in High Technology", Lessons from MIT and Beyond, Oxford et al.

Roberts, E.B., 1991b, "High Stakes for High-Tech Entrepreneurs: Understanding Venture Capital Decision Making", in: Sloan Management Review, Winter, pp. 9-20.

Rothwell, R., 1985, "Venture Finance, Small Firms and Public Policy in the UK", in: Research Policy, Vol. 14, pp. 253-265.

Rüschen, T. 1990, "Consulting-Banking", Wiesbaden.

Sahal, D., 1983, "Technology, Productivity and Industry Structure", in: Technological Forecasting and Social Change, Vol. 24, pp. 1-13.

Sahlman, W. A., 1989, "Aspects of Financial Contracting in Venture Capital", Journal of Applied Corporate Finance, pp. 23-36.

Sahlman, W. A., 1990, "The Structure and Governance of Venture-Capital Organizations", Journal of Financial Economics, Volume 27, No. 2., Oct., pp. 473-525.

Sahlman, W., 1992, "Insights from the Venture Capital Industry", Harvard Business School, mimeo.

Sahlman, W., Stevenson, H., 1987, "Capital Market Myopia", Journal of Business Venturing, 1, pp. 3-23.

Sapienza, H.J., 1992, "When Do Venture Capitalists Add Value?", in: Journal of Business Venturing, Vol. 7, pp. 9-27.

Sapienza, H.J., Gupta, A.K., 1989, "Pursuit of Innovation by New Ventures and its Effects on Venture Capitalist - Entrepreneur Relations", in: Brockhaus, R.H. et al. (Eds.): Frontiers of Entrepreneurship Research, Wellesley, pp. 304-317.

Sapienza, H.J., Manigart, S., Herron, L., 1992, "Venture Capitalists' Involvement", in Portfolio Companies: A Study of 221 Portfolio Companies in four Countries, Paper, Babson Entrepreneurship Conference.

Sapienza, H.J., Timmons, J.A., 1989a, "Launching and Building Entrepreneurial Companies: Do the Venture Capitalists Add Value?", in: Brockhaus, R.H. et al. (Eds.): Frontiers of Entrepreneurship Research, Wellesley, pp. 245-257.

Sapienza, H.J., Timmons, J.A., 1989b, "The Roles of Venture Capitalists in New Ventures: What Determines Their Importance?" in: Academy of Management; Best Papers Proceedings, pp. 74-78.

Scherer, F., 1991, "Changing Perspectives on the Firm Size Problem," in Innovation and Technological Change: An International Comparison, ed. Z. Acs and D. Audretsch, University of Michigan Press, Ann Arbor, MI, 24-28.

Schmidt, R.H., 1988, "Venture Capital in Deutschland - Ein Problem der Qualität?", in: Die Bank, Nr. 4, S. 184-186.

Schramm, B., 1988, "Finanzierung nicht emissionsfähiger mittelständischer Unternehmen", in: Christians, Hrsg.: Finanzierungshandbuch, 2. Aufl., Wiesbaden, S. 563-576.

Schröder, C., 1992, "Strategien und Management von Beteiligungsgesellschaften: ein Einblick in Organisationsstrukturen und Entscheidungsprozesse von institutionellen Eigenkapitalinvestoren", Baden-Baden.

Schütt, F.H., 1993, "Deutsche Bürgschaftsbanken in der Bewährung, in: Sparkasse, Jg. 110, Nr. 10, S. 465-467.

Soja, T.A., Reyes, J., 1989, "Investment Benchmarks", Venture Capital Journal, pp. 118.

Stedler, H.R., 1993, "Beteiligungskapital im bankbetrieblichen Leistungsangebot", in: Die Bank, Nr. 6, S. 347-351.

Swoboda, P., Zecher, J., 1985, "Unternehmensberatung und Risikokapitalbildung, in: Betriebswirtschaftliche Forschung und Praxis, Jg. 37, Nr. 5, S. 402-420.

Thackray, J., 1983, "The Institutionalization of Venture Capital", Institutional Investor, August, pp. 73-76.

Weichert, R., 1987, "Probleme des Risikokapitalmarktes in der Bundesrepublik", Tübingen.

Wetzel, W.E., 1983, "Angels and Informal Risk Capital", in: Sloan Management Review, Vol. 24, Summer, S. 23-34.

Wetzel, W.E., 1987, "The Informal Venture Capital Market. In: Vesper et al. (Eds): Frontiers of Entrepreneurship Research, Wellesley, pp. 412-428.

Wirtschaftswoche, 1995: "Reich und berühmt", Nr. 36, S. 78-89.

Workshop '83, 1983, "Venture Capital für junge Technologieunternehmen" (organisiert vom VDI-Technologienzetrum Berlin), München.

Wupperfeld, U., 1993, "Mißerfolgsfaktoren junger Technologieunternehmen", Karlsruhe.

Wupperfeld, U., 1994, "Strategien und Management von Beteiligungsgesellschaften im deutschen Seed-Capital-Markt - Ergebnisse einer empirischen Untersuchung von 33 Beteiligungsgesellschaften und Banken", Arbeitspapier des ISI, Karlsruhe.

Yin, R.K., 1989, "Case Study Research Design and Methods", London, Sage Publications.

Zigmund, W.G., 1991, "Business Research Methods", London, Dryden Press.

TECHNOLOGY, INNOVATION and POLICY

Series of the Fraunhofer Institute
for Systems and Innovation Research (ISI)

Volume 1:
Kerstin Cuhls, Terutaka Kuwahara
Outlook for Japanese and German
Future Technology
1994. ISBN 3-7908-0800-8

Volume 2:
Guido Reger, Stefan Kuhlmann
European Technology Policy
in Germany
1995. ISBN 3-7908-0826-1

Volume 3:
Guido Reger, Ulrich Schmoch (Eds.)
Organisation of Science and Technology
at the Watershed
1996. ISBN 3-7908-0910-1

Springer
and the
environment

At Springer we firmly believe that an international science publisher has a special obligation to the environment, and our corporate policies consistently reflect this conviction.

We also expect our business partners – paper mills, printers, packaging manufacturers, etc. – to commit themselves to using materials and production processes that do not harm the environment. The paper in this book is made from low- or no-chlorine pulp and is acid free, in conformance with international standards for paper permanency.

 Springer